NORTH CAROLINA TODAY

North Carolina TODAY

By

S. H. HOBBS, Jr.

and

MARJORIE N. BOND

Aids to Learning

By MARY SUE BEAM FONVILLE

THE UNIVERSITY OF NORTH CAROLINA PRESS

CHAPEL HILL

MAPS AND CHARTS BY CAROLYN McCARTHY BOLT

SKETCHES BY RUTH HINES TEMPLE

PRINTED IN THE UNITED STATES OF AMERICA

Foreword

READERS may wonder at first why this book is called *North Carolina Today*. "Why talk about today?" they may ask. "By tomorrow, the day which is now today will be yesterday." So perhaps we should explain at the very beginning that we are using the word *today* in a very general sense. We are talking about North Carolina as it is nowadays, at the present time.

In the life of a state or of a nation, the present time may be much longer than it is for any person. In the lifetime of North Carolina, for example, today may cover a number of years instead of only twenty-four hours. What is true about the state in 1950 may have been true in 1940, and it may still be true in 1960 or later.

This *today* we are talking about sometimes reaches back into the past. So a look backward often makes the present clearer to us. Many facts about North Carolina nowadays mean very little unless we know something of how they came to be.

What happens today is also a part of tomorrow. So we talk sometimes about what the state may be like in the future. And we try to see how the things which we do today may make North Carolina better tomorrow.

This book is written especially for the younger people of North Carolina, to help them understand the state in which they live.

Part I asks the question, "What is wealth?" And it describes the five kinds of wealth which we find in the world about us: natural wealth, human wealth, technological wealth, capital wealth, and institutional wealth.

In Part II we see in a general way how much of each kind of wealth we have in North Carolina.

Part III describes how we are using our natural wealth in farming, and points out some of the ways in which our resources for agriculture might be put to better use.

Part IV describes important manufacturing carried on in the state, and suggests that we could use more of our resources for industry.

In Part V we look more closely at some of the institutional wealth of the state and discover some of our resources in government, churches, schools, colleges, and other institutions.

In Part VI we see that the future of North Carolina depends partly on ourselves and how we use our resources, partly on how people live and work in other regions and other countries. To make North Carolina a balanced state is the job ahead of young people today. They are the ones who will shape the world tomorrow.

The maps, charts, sketches, and pictures have all been carefully planned to help the reader understand the book. As further aids to learning there are suggestions at the ends of the chapters and of each part. At the end of the book is a reading list for teachers.

Our discussions are based on the most recent figures available. Some of these are for the years 1945 or 1946. Some are for 1940. By 1955 or 1960, these figures may be entirely different, or they may be very much the same. No one knows how long *today* will last in North Carolina or any other state.

If the figures do not remain about the same, which way

will they change? And will the general picture of the state change with them? These are questions which only the future can answer.

Although we have read and reread all the chapters very carefully, trying to make sure that we have said things clearly and correctly, there may be errors in the book. If there are, we alone are to blame for them, and we will be grateful to anyone who points them out to us.

A great many people have helped, directly or indirectly, with this book. Some of them lived many years ago, but they helped us by the work they did—the books they wrote, the records they kept, the things they planned. Others are people living in North Carolina today who have given us special information or have read all or part of the book for us. Many of these are personal friends. Others we know only through their willingness to give us time and help when we asked for it. We are grateful to them all.

It is impossible to name everyone who has helped us in one way or another, but we wish especially to thank the following:

Professors J. C. Andrews, Gordon W. Blackwell, Richmond P. Bond, Franklin C. Erickson, Harriet L. Herring, Harold D. Meyer, Howard W. Odum, Rupert B. Vance, and Paul W. Wager of the University of North Carolina.

Professors L. D. Baver, G. W. Forster, C. Horace Hamilton, William D. Lee, and I. O. Schaub, and Mr. Frank Jeter of the North Carolina State College of Agriculture and Engineering of the University of North Carolina.

Superintendent Clyde A. Erwin, Miss Hattie S. Parrott, Miss Julia Wetherington, and other members of the State Department of Public Instruction.

Mr. W. Kerr Scott and members of the State Department of Agriculture; Mr. Forrest Shuford and members of the State Department of Labor; Dr. Ellen Winston and members of the State Department of Public Welfare; Miss Marjorie Beal of the North Carolina Library Commission; Mr. C. C. Crittenden of the State Department of Archives and History; members of the State Board of Health; Mr. Paul Kelly and Mr. Bill Sharpe of the State Department of Conservation and Development. Mr. Sharpe and his staff lent a number of photographs from which many of the pictures for the book were selected.

Mr. O. J. Holler of Union Mills; Mr. Hugh MacRae of Wilmington; Mr. Don S. Matheson of Hillsboro; Mr. Wade H. Paschal of Siler City; Mr. Thomas J. Pearsall of Rocky Mount; Mr. William T. Polk of Greensboro.

Mrs. Mary Sue Beam Fonville of Raleigh, who made many valuable suggestions and prepared the Aids to Learning.

Mr. W. T. Couch of Chicago, Illinois, formerly Director of The University of North Carolina Press; Professor H. Arnold Perry of University, Alabama, formerly in the State Department of Public Instruction.

S. H. H., Jr.
M. N. B.

ACKNOWLEDGMENTS

The authors and publisher wish to thank the following persons and organizations for the photographs used in the book: Agricultural Adjustment Administration, 167, 209; American Forest Products Industries, 70 (center), 275, 276; Albert Barden, 45 (right), 157, 245 (top); Frank Clodfelter, 2, 23 (bottom), 39, 43, 144, 271; Department of Conservation and Development, 5, 8, 20, 23 (top), 35, 36, 42, 44 (right), 45 (left), 50, 52, 53, 63, 65, 68, 75 (top), 91, 94, 95, 97, 120, 150, 152, 155, 156, 169, 175, 177, 193, 194, 197, 211, 218, 223, 234, 235 (right), 238, 239, 240, 251, 252, 253, 254, 255, 256, 273, 282, 283, 308, 322, 328, 340, 349, 350, 353, 377, 379, 380, 381, 386, 402; Durham Community Fund, 357, 361, 362; Durham Morning Herald (Photographs by Cooper), 90, 286, 288, 292, 295, 367, 371; Don R. Eckelberry from National Audubon Society, 60; Farm Security Administration, 101, 118, 126, 138, 146, 149, 213, 335; Elliot Lyman Fisher, 245 (bottom); Hugh MacRae, 183; Kenneth Ness, 104; North Carolina College, 333; North Carolina Symphony Orchestra, 102; Rockingham County Library, 346; Soil Conservation Service, 29, 70 (bottom), 75 (center); State College of Agriculture and Engineering of the University of North Carolina, 89; State Department of Health (Oral Hygiene), 373; Tennessee Valley Authority, 30; United States Department of Agriculture, 27, 179, 216, 235 (left), 281; United States Department of the Interior (National Park Service), 59; United States Forest Service, 13, 44 (left), 51, 56, 57, 70 (top), 75 (bottom); Bayard Wootten, 172, 229, 230.

CONTENTS

I. WHAT IS WEALTH?

1. *The New Found Land*
2. *The Wealth of the World*

Black bears find their way through trees and underbrush.

1. *The New Found Land*

THEY wanted gold—the Englishmen who first came to live on the shores of this new found land.

They were Raleigh's colonists, who sailed from England in 1585 with Ralph Lane as their governor, and they knew well enough that the land was good. Barlowe and Amadas and other men had come over a year earlier, in 1584, to look around and find a place for the colony to settle. And hadn't they gone home reporting that the soil was "the most plentifull, sweete, fruitfull and wholsome of all the worlde"? *

The summer was so long and so hot, Barlowe said, that in one season the Indians could grow three or four crops of corn. They also planted melons, peas, beans, and a strange vegetable known as openauk. This was "a kinde of roote of round forme, some of the bignesse of Walnuts, some farre greater," which the English called potatoes.

The inlets and bays teemed with fish. Black bears and fat deer and rabbits followed narrow trails through the underbrush between cedars and cypress trees, gums and oaks. And there were forests of longleaf pines.

* Some of the phrases quoted in this book were written when English words were not always spelled the way they are today.

Barlowe said they found grapes growing thick and sweet-smelling to the water's edge. And they saw a heron rise from a tangle of branches and flap its wings against a July sky that was cloudless and blue.

Coppery black turkeys hunted for food in the swamps and river bottoms. Mourning doves called to each other—"Ah, woe, woe, woe"—and bobwhites wandered into small clearings in search of ripening seeds.

Barlowe also reported how he and the other men had traded with the Indians. "We exchanged our tinne dish for twentie skinnes, woorth twentie Crownes or twentie Nobles," he wrote, "and a copper kettle for fiftie skins woorth fiftie Crownes." That was clever trading. A man could make money that way.

The wealth of the land lay ready for the taking. But the Englishmen under Governor Lane did not come to the new world to make their fortunes by farming or by trading with the Indians. They were here for something else.

They wanted precious stones, silver, and gold. Especially gold. The Spaniards had been bringing home gold from the land to the south. It was making Spain rich, rich enough to build more ships to carry home more gold, rich enough to fight the English or anyone else who wanted the gold for themselves.

Barlowe and Amadas had looked for gold among the savages, during the few weeks when they were exploring the coast in search of a good place for a permanent colony. They had seen Granganimeo, brother of a chief, wearing on his head a wide disk of gold or copper. The metal was so dark that Barlowe could not be sure which it was.

The Indian's wife had a band of white coral around her forehead, and in her ears she wore "bracelets of pearles hanging downe to her middle . . . of the bignes of good

A heron, rising slowly from a tangle of branches flaps its wings against the cloudless July sky.

pease." Granganimeo offered a large box of pearls in exchange for armor and a sword. But the Englishmen refused to trade. After all, what was one boxful? The settlers who were coming over would find out where the pearls were. Then they could go and get all they wanted.

It was early in July, 1585, when seven small ships, with Lane's men aboard, reached the shores of what is now North Carolina. The colony had supplies for only twenty more days. Yet the men spent no time in planting crops for food through the winter. They would get meat and corn from the Indians, they told each other.

So they built some houses and a fort on Roanoke Island, and called the colony Virginia. And then Ralph Lane and many of the men went off to look for gold.

First they went to the south 80 miles by water. Next they went to the north about 130 miles to the Chesapeake, where they found a good harbor. Then they went to the northwest about 130 miles across a broad sound, and up a river 30 miles farther. But they found no gold.

They heard of a chief who used pearls to trim his clothes and his houses and his beds and his friends. They saw an Indian with disks of silver which he said had come from a province to the north.

They heard of a strange metal that was soft and reddish yellow. Of course it might be copper. And then again it might be gold. As soon as Sir Richard Grenville came with more men and more supplies, the English would find out.

But the Indians had decided that the white men were not their friends. They planted nothing, and wanted the grain which the Indians had grown for themselves. Although the English were willing to trade for the grain, the Red Men soon learned that a man who has a tin plate is not wealthy if he has nothing to eat.

With their guns the white men could kill at a great distance. They killed Indians, and they killed the animals which Indians hunted for food. The hunting grounds were being spoiled.

No, the white men were not their friends, the Red Men decided. So they refused to trade any corn or other vegetables, or any game. And they broke the woven brushwood traps which the English had put out in the water to catch fish.

The Indians planned to set fire to the houses in the little colony, and then to kill the men as they ran out into the darkness of the night. But Governor Lane heard of this danger, and the white men killed many of the Indians instead.

That was the first of June, 1586. A few days later Sir Francis Drake arrived off the coast with his fleet, fresh from raiding the Spanish settlements to the south. He offered to give the colonists some food and ammunition, and to leave a ship with them to take them back to England later on.

At first Lane and his men said they would stay. They still hoped to find gold. But after a storm had driven many of the ships from shore, they changed their minds and sailed for home with Drake. They had been in the new world for almost a year. They left the wealth of the land behind them.

A few weeks later, Sir Richard Grenville reached Roanoke Island with food and supplies for the colony, but he was too late. The houses and the fort were deserted. Fifteen of his men stayed there, to hold the country for the Queen, and Grenville sailed away.

Raleigh was disappointed by the return of the colony, but he did not give up his plans to settle Virginia. By April of 1587 he had three ships ready to sail. John White was to be governor of this colony. And this time women and children were to go too, as well as men. Raleigh decided that men who had families to take care of would make better settlers than adventurers with nothing to think about but gold.

The ships were to stop at Roanoke Island long enough to pick up the fifteen men left there by Grenville less than

The outer banks and narrow channels along the Carolina coast were a greater barrier to early settlers than the wide ocean itself.

a year earlier. Then they were to sail north along the coast until they came to Chesapeake Bay. Ralph Lane had reported a good harbor there, and it would be safer not to live on Roanoke, where the Indians were unfriendly.

When the ships reached the new world, Governor White went ashore with a landing party in search of Grenville's men. But he found only the bones of one of them, killed by the Indians long before. Of the others there was no trace. The fort had been torn down, the houses were half lost in a tangle of melon vines, and deer stood deep in underbrush eating the ripe fruit.

The English were then ready to sail on to the Chesapeake, but Simon Ferdinando, the ship's master, refused to take them. It was late summer, he said, and he was anxious to leave for home before the September storms.

So Raleigh's second colony had to settle down to live on Roanoke Island, where the first colony had left off. The men cleared away the vines, and they repaired the houses, and soon they were digging small garden patches where they could raise some food.

When Ferdinando sailed for England, John White went with him to get supplies for his colony. But it was a long time before he was able to come back to Roanoke. At home he found England busy fighting Spain, and the Spanish Armada ready to invade England. No one was worried about helping a few colonists in a remote wilderness.

It was August 17, 1590, when John White returned to his colony in the new world. Late that day, two boatloads of men rowed towards the shore. The sky was already dark, and through the woods flashed the lights of a big fire.

The party dropped anchor, and one of the men blew a call on his trumpet. Surely the colonists must hear, must answer. Then the men sang English songs—old tunes they had been singing since they were boys. The music drifted over the water and faded into the silence of the woods. Then they called out and shouted toward the shore. No answer.

At daybreak they landed. Trees and grass were still burning, but no one was there. When the men went on to the settlement, they found the houses tumbled to the ground and almost overgrown with weeds.

The bark had been peeled from a post near the entrance to the enclosure. And a few feet above the ground someone had cut the word CROATOAN into the wood. That was all. CROATOAN. The colony was lost.

The wealth of the land was still untouched.

The next English settlers went directly to the bay of the Chesapeake, where the harbor was good. And it was this settlement farther north, around Jamestown, that became the colony—and later the state—of Virginia.

The wealth of the land to the south remained untouched for years. Then King Charles the First of England gave this land to some of his friends. And the colony was called Carolina from Carolus, the Latin form of the King's name.

So the land where the English made their first homes in the new world became Carolina. Part of it is what we know as the state of North Carolina today, with its capital named in honor of the man who tried to settle a country he never saw.

It was long after the Lost Colony disappeared that Carolina was finally settled. The outer banks and the reefs along the coast, the changing inlets and the narrow channels made early colonists hunt for an easier harbor. The shore line was a barrier greater than the wide ocean itself. When permanent settlers finally came to Carolina, they came by land.

If You Want to Learn More

1. About Barlowe's impressions of the North Carolina coast, ask your teacher to read the parts of his report found in Lefler, *North Carolina History Told by Contemporaries*, pp. 3–9.

2. About the Lost Colony, read
   a. Arnett, *The Story of North Carolina*, Chapters 4–5.
   b. Green, *The Lost Colony*.
   c. Newsome and Lefler, *The Growth of North Carolina*, pp. 18–28.

3. About the coast of North Carolina, draw a map showing the details of the coast line.

If You Want to Share What You Have Learned

1. Dramatize a meeting at which Amadas and Barlowe tell Raleigh what a wonderful land they have found.

2. Make a series of drawings or paintings to show what you think the "New Found Land" was like.

2. *The Wealth of the World*

WHAT of the gold that the English settlers were hoping to find in Carolina? There was gold here. Not very much of it, to be sure, but there was some. Men have found pieces of gold heavy enough to hold open a cabin door in summer time.

There is gold in North Carolina, just as there is gold in other parts of the world. But somehow people do not want it as much as they used to. They have found that gold is only one way of producing wealth and that there are several other ways to do this. And they have found that there are other kinds of wealth just as important as gold.

What is wealth, then, if it is something more than gold?

The wealth of the world may be divided into five general classes. These five kinds of wealth are:

1) *Natural wealth.* This means soil, climate, rivers, lakes, trees, wild plants, birds, animals, fish, and minerals. It means all the things which people find in their natural surroundings. The wealth of the world depends on our natural resources and how we use them.

2) *Human wealth.* This means people. People who use the natural wealth they find around them. People who know things, people who make things, people who do things, people who use things.

3) *Technological wealth.* This means the knowledge and skills with which people work, the way they use their hands and their minds. It is a combination of imagination, education, intelligence, and machines.

4) *Capital wealth.* This means money, or credit with which to borrow money. Since people have put a money value on gold, the mineral Lane's men were looking for might be part of our capital wealth.

5) *Institutional wealth.* This means all the organizations and institutions which people have created and which make it possible for them to have a more comfortable and more satisfactory life.

Each of these five kinds of wealth depends on all the others for complete development. One cannot be fully developed without the others.

Tom Caswell, for example, is a farmer with a hundred acres of land. But this farm is not just so much space. It is natural wealth. It is the land on which Tom can grow a crop and thus change his wealth of land into another form of wealth.

Tom has the soil, which is rich, crumbly loam full of food for his crop. He has the warmth of the sun. And he has the rains, which are building up the supply of plant food or changing it into a form easily used by the plants. Tom has the stream which curves across the bottom land. He and the children go fishing there. He has trees which stand ready to be cut for firewood or lumber. There are birds and animals for him to hunt.

All this Tom Caswell has on his hundred acres, and it is wealth, *natural wealth.* The wealth is there, whether Tom

Wealth today depends on natural resources more than on any other one thing. It depends on how we use our forests, our soils, our minerals, and everything else we find in the world about us.

Caswell is there or not. It is potential wealth, possible wealth. It is the wealth of the land.

In order for this potential natural wealth to be used, there must be people. People are human wealth. They are a kind of wealth, just as much as the gold and silver and precious stones that Governor Lane's men hunted for.

There must be people to work. "Man power," we call them now. "Hands," they are called on farms and plantations, where men and women working in the fields are just so many hands changing the natural wealth into something easily available.

Tom Caswell is a part of this human wealth, and so are his wife and children, who help with the work on the place. They help to convert the potential wealth of his hundred acres. Men and women, then, are a kind of wealth. *Human wealth.*

These men and women are worth more when they are well trained and know how to do things. Their knowledge and their skills are a part of what is called *technological wealth.* Tom Caswell has some of it. He uses a simple skill when he produces his crops. Tom's brother, Jim, who works in a hosiery mill, is part of our technological wealth. He is trained to control the machines through their complicated processes. And Lawyer Bob's younger boy, Fred Caswell, is part of our technological wealth. He went off to college, and now he works in the laboratory of a textile mill, finding out better ways to make things and do things.

That is what technological wealth really is. Not just hands, but skillful hands. And machinery. And the brains to make machinery that will work for skillful hands.

But to buy the machines, to build the factories, to buy Tom Caswell's crop, to pay Jim Caswell and Fred, a fourth kind of wealth is necessary—*capital wealth.* This is sometimes money. It is sometimes credit, which is the ability to borrow money.

Henry Caswell had to have capital in order to build the factory and the machines which use Tom's crop, Jim's skill, Fred's brains. Men have to have money in order to make money. Tom and Jim do not have much of this kind of wealth. Neither has Fred. But they need someone to have capital wealth, just as capital wealth needs them.

Men have used some of their wealth to create for themselves a fifth kind of wealth—*institutional wealth.* Schools and churches are a part of this, and town and county governments, and state and federal governments. Through these institutions men plan and work for a better way of life for everyone, a better way to use the natural wealth without wasting or destroying it.

Everyone gives some money to pay the expenses of these institutions—Tom and Jim and Fred, and Lawyer Bob, and Henry, and you, and I, and the folks next door, and the folks in the next town, and the folks on the next farm, and the folks in the next county. Everyone. Everyone shares in the plans. Everyone lives in a better way because of them. Institutional wealth means community wealth, common welfare.

These five sources of wealth depend on one another for their fulfillment. Natural wealth, human wealth, technological wealth, capital wealth, institutional wealth—each one needs the other.

Are You Sure

1. That you know the difference between the five kinds of wealth defined in this chapter?

2. That you understand the close connection between all these kinds of wealth?

3. That you can explain what is meant by *potential* wealth in contrast to actual wealth? When is fertile land potential wealth rather than actual wealth? In what sense is an uneducated but intelligent man or woman potential wealth as well as actual wealth?

Why Not

1. Start a word list in your notebook where you can put the new words and terms you learn while studying about North Carolina? Start with the different kinds of wealth described in this chapter.

2. Arrange a display of examples from your own community of the different kinds of wealth? Include both pictures and objects. If some of you take photographs as a hobby, you might make pictures instead of collecting them.

3. Begin a class book illustrating through pictures, graphs, selected clippings, and the like, the (a) wealth of your community, (b) examples of other wealth in North Carolina?

4. Discuss in class how people's ideas about wealth have changed since early times. What was wealth to the Spanish explorers? To Sir Walter Raleigh's countrymen? To the Indians? And to the first settlers? How do our present ideas about wealth differ from theirs?

II. DISCOVERING THE WEALTH OF NORTH CAROLINA

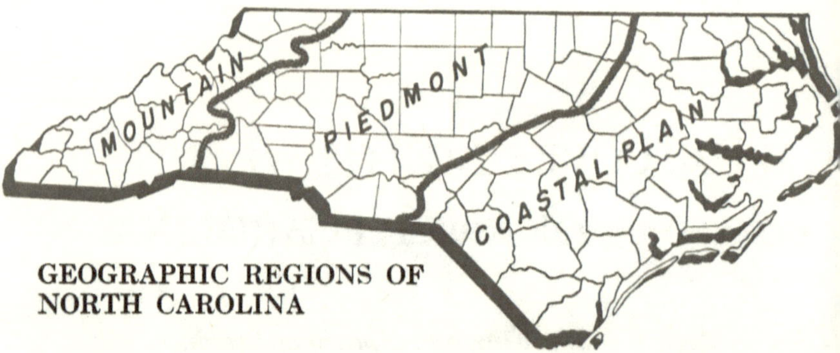

GEOGRAPHIC REGIONS OF NORTH CAROLINA

The lay of the land divides North Carolina into three great geographical areas: the Coastal Plain, the Piedmont, and the Mountain region.

3. *"Fair and Spacious Province"*

WHAT is this North Carolina, the Old North State? It is still "a fair and spacious Province," as one man described it back in 1666. But what does it look like? How does the land lie?

It appears on the maps of the United States of America with a definite and familiar outline. We all know how it looks on a map. But suppose we could see North Carolina the way we see a man standing in silhouette against the sky. What would the profile of the state look like then?

There is a map inside the front cover of this book that shows what North Carolina's profile is like. Along the coast the land is at sea level, and for a number of miles to the west it remains very low. Then it rises higher, in a broad plateau. And then it rises even higher, into mountains.

This is North Carolina's silhouette. This is what the profile of the land is like.

And how does the state look to an ambitious and curious boy like Fred Caswell's fifth cousin, Dick? Dick and some of his school friends are walking from their homes in Manteo on the coast all the way to the mountains. What do they see?

As they cross the state they cross North Carolina's

In the east this "fair and spacious Province" is flat and low, dropping off gradually until land meets water.

three great geographical areas: (1) the Coastal Plain, (2) the Piedmont, and (3) the Mountain region.

The Coastal Plain itself may be divided into two subregions of its own, the Tidewater and the Upper Coastal Plain. The boys walk through the Tidewater region first. The Tidewater is the land along the coast which is reached by the water of the tides in the rivers.

The shore line of North Carolina is cut into irregularly by the waters of ocean and sounds. The country is so flat that some of the rivers wind back and forth many times before they finally reach the sea. Wherever the land appears above water, it is low-lying and swampy. This country is known as the Tidewater. It extends inland away from the coast as far as the tide waters reach.

Where the Tidewater ends, the Upper Coastal Plain begins. Here the profile of the state rises in broad terraces, or steps. It goes up so gradually that the change is hardly noticeable. Certainly Dick Caswell and his friends, on their way to the mountains, do not know that they are doing any climbing.

When the boys come to the western margin of the Coastal Plain, they find that the level of the land changes. Rivers flowing toward the coast drop suddenly, in waterfalls or rapids, as the land falls away sharply.

Dick and his friends have reached the fall line, which is the western limit of the Coastal Plain. What is the fall line? In one way the fall line is imaginary, and in one way it is very real.

Long, long ago the Tidewater and the Coastal Plain were under water, part of the floor of the ocean. Then something happened, and the ocean floor was pushed up. It is no longer covered by water, but it is still lower than the land farther inland. The fall line is the name that has been given to the old shore line that once bordered the old coast.

In North Carolina the fall line runs northeast and southwest, almost parallel to the shore. It passes through Northampton, Halifax, Nash, Wilson, Johnston, Wake, Harnett, Lee, Moore, Montgomery, Richmond, and Anson counties. It is really more of a belt than a line.

To the west of the fall line there are more hills, and the rivers and the roads curve through rolling country. This land which lies at the foot of mountains is called Piedmont, just as in Italy there is a Piedmont from the two Italian words *piede* meaning *foot* and *monte* meaning *mountain.* As the boys from the coast walk along, they go more slowly. The road is hilly, and they see mountains ahead.

When Dick Caswell and his friends reach the Mountain region, the Land of the Sky is all around them. They are in the Carolina highlands. First they come to mountains that look blue even though the trees that cover them are green. These are the Blue Ridge mountains. Then they come to mountains which are often hidden by clouds that look like smoke. These are the Great Smokies. Between the top of the Blue Ridge and the top of the Smokies are the Carolina highlands.

Here the profile of the state juts into the sky. Forty-nine mountain peaks are more than 6,000 feet above the level of the sea, 174 peaks are between 5,000 and 6,000 feet. Here stands Mount Mitchell, the highest mountain east of the Mississippi. Here are Clingman's Dome, Potato Knob, Cattail Peak, Mt. Guyot, Balsam Cone, the Black Brothers, Mt. Gibbes, and many others almost as high. Here is old Grandfather Mountain—not so tall as some of the others, but a great favorite with Tar Heels.

By this time Dick Caswell and his friends have had quite a walk. From one end of the state to the other—from the shore to the Tennessee line—North Carolina measures 503 miles. The boys are a long way from home.

They have had quite a climb, too. Starting on the coast at sea level, they have climbed to mountains which are 6,000 feet or more above the level of the sea. They have walked up the profile of the state, and have seen the three great geographical areas of North Carolina.

You'll Want to Know the Answers to These Questions

1. What is meant by the term "fair and spacious province?" How did it originate?

2. What is meant by saying that the state rises in broad terraces, like wide steps, from east to west?

In the west, mountain ridges disappear into the distance, and the land which is Carolina rises until it meets the clouds.

3. How do the different sections of North Carolina (Coastal Plain, Piedmont, and Mountain) compare in size and other characteristics with similar areas in her neighbor states?

4. What is meant by the geologic history of our state, and how does it help explain why there are such differences in the three sections of North Carolina? If your school does not already have a copy of Bryson: *The Story of the Geologic Making of North Carolina*, perhaps your teacher can secure this free pamphlet from the Department of Conservation and Development in Raleigh.

You'll Want To

1. Read what your science books and reference books say about changes in the earth's surface. Then you will better understand what has happened to North Carolina through the centuries. See:
   a. Carroll, *Understanding Our World*, Unit XI.
   b. Smith and Trafton, *Enjoying Science*, Unit X.
   c. Wetherington and others, *Science in the Elementary School*, Unit II, "The Earth Has Had Many Periods of Change," p. 85.

2. Read Chapter 2 in Rowe, *Discovering North Carolina.*

3. Collect or make pictures showing scenes typical of each section.

4. On an outline map of North Carolina show through which counties the "fall line" runs.

4. *Soils*

TO MOST people, soils seem more or less alike. But to farmers, in North Carolina and everywhere else, soil is something special. A man has to understand it. He has to know what it's like, what it can grow, and what it might grow if it had certain things done to it.

If Tom Caswell is thinking about buying the ten-acre field next to his farm, he goes over, leans down, and takes up a handful of the soil. He feels of it, he presses it together, he watches to see if it breaks apart in his fingers or remains in solid lumps.

If the soil is sticky when it is wet, Tom knows he will have a harder time cultivating it. If the soil is coarse or sandy, it will be easy to cultivate. Between the wet sticky clay and the light sandy soil are the loams—heavy, medium, and light.

There are other things Tom wants to know.

Is there enough rain? And can the soil hold moisture? Tom knows that the answer to this question depends on whether the soil contains humus. Humus is formed by decaying leaves, stems, and other organic matter. It gives life to the soil and helps to keep it moist.

Tom looks at the field carefully to see how much it slopes. Even though the soil has humus to hold moisture, the land may be so hilly that most of the rain runs off before it soaks in. Wherever plowed slopes are steep, the fertile topsoil will be washed away by the rains. Soon gullies will appear in the fields where the water makes its own paths. This is erosion, and Tom knows its dangers. He also knows that if he is careful in his plowing and his planting, he can use even a hilly field. He knows how to control erosion.

Tom has other questions he needs to answer. Is the temperature suitable? Is the soil deep enough to let the roots grow as they should? Does it have enough air for the roots?

And is the soil acid or alkaline? Tom never used to pay any attention to this. Then the county agent showed him that the acidity of the soil is important. So the agent comes over and takes soil samples from the field Tom is thinking about buying. He makes a quick test for acidity right there in the field. Then he packs some soil samples in special boxes and sends them to the Soil Testing Service of the North Carolina Department of Agriculture in Raleigh.

When Tom gets the report, he knows whether the soil is acid or alkaline, "sour" or "sweet." Some crops need one kind of soil, some crops need another. The soil may be too acid for the crop Tom was planning to grow in that field. But he can reduce the acidity with lime, and the report will tell him about this.

The report also answers one of Tom's other questions: Does the soil have the necessary plant foods—especially nitrogen, phosphoric acid, and potash? And are they in a form that the plants can use? If his crop needs a food which is not there, Tom can add it. This is important, because plants cannot grow well without proper food, any more than people can.

These are a few of the questions Tom needs to ask about that ten-acre field before he buys it. When we learn these things about the whole state, we know something about North Carolina's natural wealth in soils.

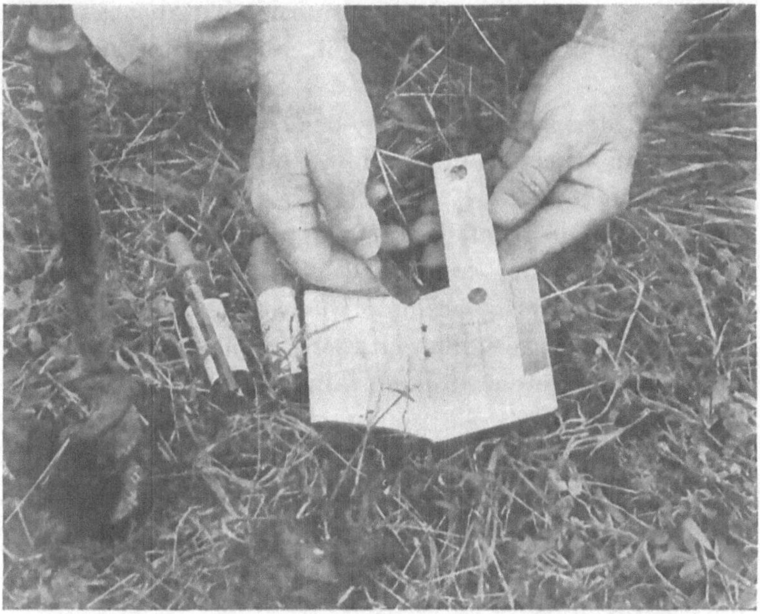

The county agent takes samples of the soil and makes a quick test for acidity right in the field.

Anyone who digs a hole several feet into the ground will notice that the soil he takes out of the hole is not all the same. He is digging through layers of different kinds of soil.

These layers differ in the amount of plant and animal material they contain. They also differ in the size and amount of the solid material they contain—whether this material is mostly sand, gravel, stones, or clay. Various kinds of soil look different to the eyes and feel different to the fingers.

The first layer of soil on the surface of the ground is the topsoil. Topsoil generally contains more plant and animal materials than the other layers. This is the layer which is used most for growing crops.

Below the topsoil is a layer of subsoil. This has less plant and animal material than topsoil, and it often has more clay.

Below the subsoil there may be a layer of rock. It is this rock which breaks and crumbles and becomes soil. All soil was once rock.

Soil is made from decaying organic matter and from rocks which during countless centuries have been broken into tiny bits by the action of the weather. Why are soils different? There are different kinds of soils (1) because the parent rocks were of different kinds to begin with, (2) because the rocks broke up under different conditions of moisture and temperature, and (3) because the soils have been covered with different kinds of plant materials.

Because of the changing profile and the variety of weather in North Carolina, soils vary greatly from one part of the state to another.

When young Dick Caswell of Manteo walked toward the west, he was watching the profile of the state, and he probably saw little else on his trip. But on his way home he noticed the difference in soils and crops as he walked from the Highlands to the Tidewater.

Dick had worked on Tom Caswell's farm for two summers, and had stood around on Sundays listening to the men talk of land and politics. He knew what it meant when he saw fields of barley instead of corn, and when he saw tobacco or cotton and then sweet potatoes and peanuts. Crops changed with changing soils and location.

In the Carolina Mountains are some of the most fertile

soils of the state. On the high places rich loams are held in thick layers by the carpet of fallen leaves and by the roots of grass and ancient trees. In the valleys the soil is often even more fertile than on the mountain slopes themselves. In the rolling foothills the soil is generally not so fertile. It is thin sandy loam or clay loam like the soil of the Piedmont, and is frequently gullied.

Because leaves and other organic matter are plentiful, and because they decay slowly in the low temperatures of the mountains, the soil there contains humus. For most of their crops, however, the farmers need to add phosphoric acid and nitrogen to the plant foods already present.

Even though there are good loams, with plenty of humus and some plant foods, much of the land in the Carolina Highlands cannot be farmed. If the mountains were cleared, the rich soil would soon be washed away from the steep slopes. So the farmers leave much of the soil untouched. Or they use it for crops which cover the fields and do not have to be cultivated. Then the soil is held in place.

This is a profile picture of one kind of soil in North Carolina. The thin dark layer at the top is the topsoil. Below that is the subsoil. And at the bottom is a layer of clay.

Piedmont soils are sandy loams and red clay loams. They are stiffer than those of the Coastal Plain and in general are harder to cultivate. The sloping land soon carries away the rain, and the fields dry out rapidly.

The soil in the Piedmont often lacks the humus and some of the plant food necessary for good crops. Yet in other ways much of the land is well suited to agriculture. So the farmer is learning to supply what the soil needs, and to plant what the soil will grow.

In North Carolina's Upper Coastal Plain the farmer finds the sandy loams easy to work. The soil there does not have much humus, and rains disappear quickly into the countless streams threading their way across this part of the state. Potash is in a form which plants cannot readily use, and the supply of nitrogen and phosphoric acid is low.

To this farmer and county agent, seeing is believing. Before this field was planted with rye and lespedeza, it was fertilized with superphosphate and limestone—all except a narrow strip down the middle. It is easy to see what the right kind of fertilizer did for the crop.

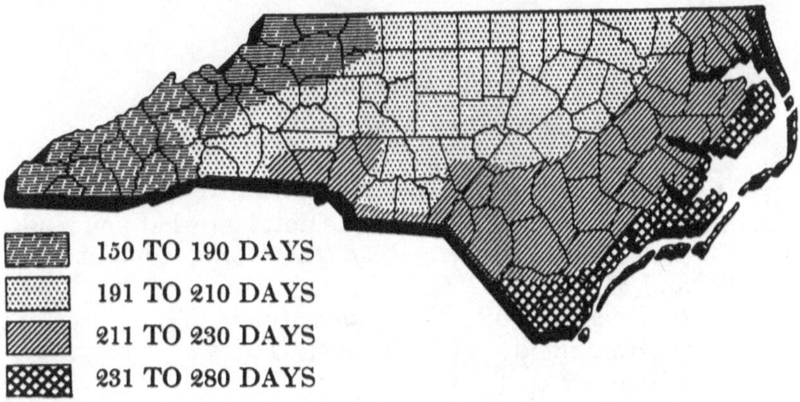

AVERAGE LENGTH OF THE GROWING SEASON IN NORTH CAROLINA

The farmers know this, and they are always adding chemical plant foods to the soil. Every year more fertilizer is used there than anywhere else in the United States.

The Tidewater has some of the best and some of the poorest soils in North Carolina. Acres and acres of sand are useless. Much of the land is so flat and so low that the water runs off very slowly. Around the bays and swamps the soil may be rich, but it cannot be used for farming until it is drained. When the land is high enough to be dry, the sandy loams are often naturally fertile.

Ever since 1900, agricultural chemists have been studying North Carolina's soils. Their tests show a general lack of humus and of available plant foods all over the state.

These tests also show that North Carolina's soils can be made better. Plant foods and humus can be increased. Fertilizer can be added. The washing away of soil can be controlled. North Carolina's soils are soon improved if we treat them well.

But this takes work. And it takes money. How is it, then, that agriculture is such an important source of wealth in North Carolina? Why does it pay? The answer lies in something that cannot be changed, no matter how much anyone might wish to do so. The answer is climate.

Be Sure to Learn More About Soil

1. Read about soil in the references listed below, or in other books or pamphlets. Take notes on what you read.
   a. Carroll, *Understanding Our World*, Chapters 50 and 51.
   b. Deusing, *Soil, Water, and Man.*
   c. Hammonds and Woods, *Today's Agriculture*, Chapter 23.
   d. Kelly and Baskerville, *The Story of Conservation in North Carolina*, Chapter 2.
   e. Melrose and others, *Nature's Bank—The Soil.*
   f. Parker, *Soil.*
   g. Watkins and Perry, *Understanding Science*, Unit II.

2. Collect in glass bottles or jars samples of soil from your school grounds, your yard, cultivated fields, woodland, roadside cut, creek bed, etc. Label these carefully to show where each came from. Examine your samples by looking at them closely under a magnifying glass or microscope. Rub the soil between your fingers. How are the samples alike? How different? How can you explain the differences? Be sure to record your findings.

3. Make a trip to study an unprotected roadside cut or a place being excavated. Study the "soil profile"; that is, study the vertical section of the ground where the different layers of soil can be seen. How deep is the topsoil? What does the subsoil look like?

4. Test the soil by the litmus paper test in *Science for the Elementary School*, Unit IX, "Conservation," p. 99.

5. Add these words to your notebook list and to your vocabulary: humus, topsoil, subsoil, organic matter, erosion, phosphoric acid, alkaline, fertility, soil profile, potash, and nitrogen.

5. *Climate*

"EVERYBODY talks about the weather, but nobody does anything about it," Mark Twain is supposed to have said. Fortunately for the farmer, not a great deal needs to be done about the weather in North Carolina. He probably complains about it as much as anyone else does. And there is hardly a season of the year which does not seem too hot or too cold or too wet or too dry. Yet year in and year out, the weather in North Carolina is the farmer's friend.

To know what kind of weather—or climate—North Carolina has, it is necessary to know the answers to a number of questions.

How much does the sun shine during a year's time? How much rain falls? Does most of the rain come within a few months, so there is a rainy season and then a dry season? Or does rain fall regularly throughout the year? What is the coldest temperature the thermometer ever shows, what is the hottest, and what is the average?

How late in the spring does a killing frost generally come, and how early in the fall? The days between these two dates make up the growing season, when plants are able to grow. What of the winds that blow, and the winds that don't blow? They have something to do with the weather. These facts are all important in describing the climate of a place.

The climate of North Carolina is generally regular and uniform, without extremes. There are times when Tom Caswell thinks a "dry spell" will never end. And there are other times when it seems as if the rain will wash away every seed or drown every plant on the place. In some years a late spring frost kills every seedling Tom has above ground. But these times are the exception, not the rule. And climate is described according to what has happened over a period of many years, not in one year or in five years or in twenty-five years.

How do we know what the weather was like ten or twenty years ago? Records of North Carolina weather are kept in many parts of the state. Weather stations are scattered from Murphy and Bryson City and the top of Mount Mitchell all the way to Wash Woods, Hatteras, Beaufort, and Southport, and at many places in between. Their records tell the story of North Carolina's climate.

In the southeastern part of the state, the average yearly temperature is 63°. The summers are hot, but there is generally a cooling breeze. The weather is so mild that some plants of the warm tropics can grow there. That is why palmetto trees appear on Smith Island at the mouth of the Cape Fear River, spreading their tops, like plumes, against the gray-green of live oaks.

Summer is short in the high land of the Blue Ridge and the Smokies. March sees plenty of snow. It is not until late in spring that the sprouting oak leaves are as big as squirrel's ears. And it is early in fall when the katydids begin arguing back and forth from hill to mountain and back to hill again.

In October the first killing frost reaches down into the

valleys. There is not time enough for some kinds of crops to mature because the growing season is so short. Where the land is from 3,000 to 6,000 feet or more above the level of the sea, the weather is cool. The average yearly temperature is about 50°, and even in summer it is not very high.

Most of North Carolina has a yearly average of about 57° to 60°. In summer there are a few days when the temperature is over 100°, and many times it is 90° and over. It is so hot that Tom Caswell's cows stand in the shade of a persimmon tree switching flies with their tails instead of hunting for tufts of grass in the sunny field.

In winter the temperature goes as low as 10° or 12°, sometimes even below zero, and Tom asks the mail carrier, "Well, is it cold enough for you?" But these extremes of heat and cold do not generally last very long.

In July, the state's normal temperature is not much higher than the temperature in New York. In January, the state's normal temperature is from 15° to 20° warmer than New York's. These warmer winters help the North Carolina farmers. Tom Caswell doesn't need as

On the top of Mount Mitchell a watcher reads the instruments which record the weather. North Carolina has more than one hundred stations reporting temperatures, winds, and rainfall.

Smith Island, in the southeastern part of the state, has such a mild climate that palmetto trees and other plants of the tropics grow there.

large a woodpile as the farmer in the North. He can spend his time doing other things besides cutting wood.

On days when the northern farmer is snowbound, the weather may be warm enough for Tom to mend fences, burn underbrush, and do many other chores. He can grow turnip greens and collards and other hardy winter vegetables. These help to give his family the kind of food they need, and also bring in extra money when he takes a basketful to the store.

In some ways cold winters help the northern farmer. His soil is made better by the freezing and thawing of the small rocks. Freezing weather improves the physical character of the soil and makes it easier to plow and work. It also kills many of the insects which are harmful to crops. The melting snows of northern winters run off the soil slowly, without washing it away.

The warmer North Carolina winters have some other advantages over the colder weather farther north. In the winter Tom Caswell can grow a cover crop which will keep the rain from washing away the soil. This cover crop will also add necessary humus and plant food when it is plowed under.

Because the winters are short, Tom can start plowing his fields early. North Carolina's weather stations keep a record of the last day in spring on which there is a killing frost. They also keep a record of the first day in the fall when there is a killing frost. These records, kept over a number of years, help the farmer know how soon in the spring it is safe for him to plant his different crops.

The months between the last killing frost in the spring and the first killing frost in the fall make up the growing season. In North Carolina this season varies from east to west, just as the temperature varies.

Parts of the coast have a yearly average of 280 days without frost, but they do not have much soil which can be used for crops. The growing season on the coast for the land under cultivation is about 240 days—50 to 70 days longer than it is in some parts of the mountains. In most parts of the state it is 210 days or longer. Nowhere in North Carolina is there a really short growing season.

The mild climate makes life easier for everyone in the state. Many people who live in colder sections of the country come to North Carolina for the winter because they want to get away from the cold weather and ice and snow.

And what of the rainfall in North Carolina? How much rain do we generally have?

From 1887 to 1941, the yearly rainfall in the state averaged 49.72 inches. Of course in some years the average has been much lower than this. In 1925 it was only 37.32 inches, the lowest it has ever been since records were kept. And for some years it has been greater—62.66 inches in 1901. In 1936, when most of the country was having very dry weather, North Carolina had plenty of rain—60.27 inches.

North Carolina's average of about 50 inches of rain is only half of the story. The other half is just as important. It is the fact that our rain is not limited to a few months of the year. Fifty inches of rain, if it all fell during a few brief periods, would do more harm than good to crops and soils.

In North Carolina it rains fairly often, and usually not a great deal at one time. The largest amount of rain generally falls during July and August, when the crops need it most. Since little of the soil in the state has enough humus to hold moisture, these frequent rains are needed to keep the ground moist.

In the mountains of North Carolina the winters are often cold, and there is usually plenty of snow.

You Can Learn About Climate This Way

1. Observe weather conditions from day to day for three or four weeks. Fix a chart on which you keep a record of the weather. This chart might show date; temperature; rain, cloudy, or sunny; wind; moisture.

2. Watch the local papers for a summary of weather conditions. Many newspapers print a short column about the weather.

3. Read what your science books say about climate. And read and report on other interesting books, such as

Fair and Warmer by Joseph Gaer, or *Weather*, by Gayle Pickwell.

4. Write a summary of the good and bad points about North Carolina weather. Imagine you are preparing a pamphlet to be sent to people in other states. Tell both the yearly average and the range of temperature. Point out the plentiful rainfall and the fairly even distribution of that rainfall. And do emphasize the long growing season.

5. Make a study visit to your nearest airport that has weather recording apparatus.

6. Invite the meteorologist from your local weather bureau station to talk to your class.

7. List the ways in which climate has affected the people in the different regions of the state.

6. *Water*

RAIN is a part of the state's wealth long after it has fallen on fields and roads and streams and forests. North Carolina has water for plants, animals, and men; water as a source of food; water for people's pleasure and health; water for transportation; water for power.

Water is necessary in order for men to live, but as long as there is plenty of it, everyone takes it for granted. Tom Caswell pumps as much as he wants, and Jim and Fred turn on their faucets as often as they like. Yet there have been a few times in North Carolina when country wells went dry and the supply of water for towns and cities grew less and less.

When water is scarce, schools have to close. Sometimes factories have to stop work. People living in towns are not allowed to use water whenever they wish. Everywhere there is danger of fire, and sometimes the smoke of burning forests lies heavy over the land. Then everyone begins to realize how important water is.

Water is a source of food. Along the coast, in the brackish and fresh-water sounds like Currituck and Albemarle, in the salt-water sounds like Pamlico, and in the waters of the measureless Atlantic, all kinds of fish are plentiful.

Fishermen in small boats spread a large net in the water, trapping quantities of menhaden. Then a larger boat comes alongside and takes the fish aboard.

There are white shad, striped bass, black bass, perch, mullet, spot, croaker, drum, bluefish, sheepshead, hogfish, squeteague, Spanish mackerel, whiting, menhaden, and many, many others. Altogether almost 350 different species of fish are found in North Carolina waters. There are oysters, too, and clams, escallops, shrimps, crabs, and diamond-back terrapins.

Many men living on the coast make a business of fishing. During the year they catch tons of edible fish and other sea foods, which are packed in ice and shipped to inland towns. Some kinds of fish which are not good for food, like the menhaden, are used for fertilizer and oil.

Fishing is a sport, not a business, for many men who live in North Carolina and for others who come here from all over the country. They go to the coast, rent a boat, anchor for a morning in the uneven swell of an inlet, and fish for bluefish. Or they take their harness and heavy tackle and try for the biggest channel bass of the season. And sometimes, in a large boat, they go "outside" and fish in the ocean.

Lawyer Bob often leaves his office and spends two or three days in the mountains, casting his line into streams for small-mouth bass or speckled trout, trying his luck in pools and lakes.

Now and then Tom Caswell's boy, Randy, digs a canful of worms and walks down to the river's edge carrying a sassafras pole he'd cut from the woods. His dog Frisky follows close behind. They climb into the boat, and Randy pushes away from the bank into deeper water.

All through the summer afternoon he fishes. If he waits until the cool of the evening, when the fish rise to feed in the shadows, he'll probably have plenty to take home for supper.

Sometimes Frances, his older sister, decides to join him. And if she does, she may have a bigger catch than Randy's when she walks up the road to the house.

For years men fished in North Carolina waters without

A boy with his dog for company can fish through a whole summer afternoon.

any restrictions. They often caught fish that were very small and fish that were about to spawn. Then some kinds of fish, like shad, became very scarce, and men realized that fishing had to be done with a plan for the future.

So now there are laws which regulate fishing. These laws limit the seasons and places for fishing, the number and size of the fish which may be caught, and the size of the nets which commercial fishermen may use.

People enjoy other water sports in North Carolina besides fishing. They go to the coast, where they swim and play in the surf and then lie on the beach in the sun. They go to lakes and streams and swimming pools. They have many kinds of boats—rowboats and canoes, motor boats and sail boats.

Water is used for highways, too. In the early days, when the land was still covered by dense forests, there were no roads—only narrow and sometimes dangerous trails. Water was then an easy highway.

Today, across a countryside that has been cleared of

Water from North Carolina's mountains starts on its way to the coast. Sometimes it falls suddenly from one level to another. Sometimes it curves around the hills and races through the valleys.

trees, men can travel fast and far on smooth roads of gravel or macadam or concrete. Yet North Carolina is rich in its water highways. Boats can go up the rivers, often as far as the rapids which mark the fall line. The very water which makes it difficult to build roads on the curving shore and low-lying land of the coast can be used for highways.

The islands and reefs are still a barrier off the shore, just as they were in the sixteenth century when English ships were trying to find a safe harbor. But the islands which form that barrier are also a part of the state's wealth. They wall off the ocean from the sounds, and small boats and barges which would not be safe outside on high seas can travel inside across the sounds and along the inland waterways.

Long ago men learned to make water work for them. Water flowing from one level to another and falling over a wheel will turn the wheel, and that wheel will turn many others. This is energy, power. It can be used to do man's work. It can be changed into hydro-electric power.

By the time water reaches the level of the Tidewater, its progress is slow. Sometimes a river winds back and forth, meeting itself many times before it finds the ocean. Sometimes it is wide and almost motionless.

Because of its rivers, its evenly distributed rainfall, and the profile of the land, North Carolina has a great source of this power. Streams rising west or east of the Blue Ridge fall either westward or eastward with the land. In the Piedmont, the rivers drop to the fall line, draining a large area. It is there that most of the state's power sites are located.

Every year electricity is within reach of more people—people on farms as well as in towns and cities. Tom Caswell has electric lights in his house, and his wife has an electric iron and an electric refrigerator. She is saving her egg money for an electric washing machine. On the next farm John Taylor has running water pumped with an electric pump, and his wife cooks on an electric stove. Rural electrification is helping to do some of the work in the kitchens, the barns, and the sheds.

With electricity making work easier and life more pleasant, the people of North Carolina are sharing in the natural wealth of the state.

Why Not Read

1. What your science books say about water?

2. What other interesting books tell you about water—such books as *The Sea for Sam* by Reed and Bronson, and *Fresh and Briny* by Rogers and Beard?

Have You Ever Wondered

1. How important underground water is? Whether it is increasing or decreasing in North Carolina?

2. Who owns the waterways of North Carolina and of the United States? Why there are laws controlling their use?

3. Where the water supply for a town or city comes from? How is it stored, filtered, and distributed for use? Make a trip to a water plant to find in detail how this service is provided.

4. Who inspects the water supply of your school plant? Has the water been tested by your local health department, district sanitarian, or State Board of Health? How much does it cost a citizen to have his water analyzed? Do you think he gets his money's worth from an analysis?

Why Not Divide the Class into Committees to

1. Report on how North Carolina makes use of her water resources for personal and household use, pleasure, transportation, and power.

2. Make a large map of the state showing the chief waterways. Indicate with one color the course of the rivers which flow into the Atlantic Ocean, and with a different color those which finally flow into the Gulf of Mexico. Use a third color to trace the course of the Inland Waterway. Make the lines for navigable rivers wider than the others.

7. *Trees and Wild Plants*

NORTH CAROLINA'S wilderness stood rich in trees and plants long before 1584, when Barlowe reported to Walter Raleigh that "the woodes are . . . the highest and reddest Cedars of the world, . . . Pynes, Cypres, Sassaphras, . . . and many other of excellent smell and qualitie. . . ." He found "above fourteene severall sweete smelling timber trees, and . . . they have those Okes that we have, but farre greater and better."

Barlowe, of course, was writing of what he saw along the coast, and had no way of imagining the limitless forests stretching far to the west, from the palmettoes around Cape Fear to the balsams and spruce trees of the high mountains.

Today, many of these forests have been cut down. Even in the days of the colony, men began using longleaf pines to produce tar, pitch, and other naval stores. Trees that grew tall and straight were cut especially for the masts of ships. Some trees were cut for lumber. Others were cut just to get them out of the way, because men wanted to use the land on which they stood.

Many of North Carolina's original forests are gone. But the state is still rich in trees. Of the 31,193,600 acres in the state, nearly 20,000,000 acres are forest land.

Loblolly pines are common in the Coastal Plain, where they are mixed with red and white oaks and other hardwood trees. In the Tidewater swamps along the coast the southern white cedar grows, the tree we often call "juniper." Near by stands the cypress, knee-deep in water.

In the Piedmont the old forests of oak and hickory are gone—cut long ago to make way for plowed fields. And when the fields were abandoned, pines grew where the hardwood forests had been. The most common pine in the Piedmont is the shortleaf pine. The loblolly is also here. And there are still some hardwoods scattered throughout the area.

In the forests of the Mountain region chestnuts used to be plentiful, but most of them have been killed by disease. Yellow poplars grow tall and straight in the mountain coves, along with white and red oaks. And scattered here and there are hickories, maples, and other hardwoods.

Shortleaf pines, Virginia pines, and white pines find a foothold on the lower slopes of the mountains. And higher up are hemlocks, spruce, and fir—trees that can live even when the weather is very cold and the land is covered with snow.

North Carolina's hardwood forests were cut so recklessly that furniture factories in the state now have to bring much of their wood from other states and countries. But North Carolina still ranks high in the lumber industry. In some years we produce more than a billion feet

In the Tidewater swamps along the coast the cypress trees stand knee-deep in water.

of lumber, as well as many articles which are made out of wood.

And our forest resources are only partly used. Every year some of our trees can be cut for lumber and furniture. And some can be cut for wood pulp, which is made into paper and modern plastics.

Men are learning that it does not pay to cut trees recklessly, the way people used to do. Many owners of wooded land now say that modern forestry methods pay in dollars and cents.

Every year Tom Caswell cuts his own firewood from the northern slopes on his hundred-acre farm. And in winter he sometimes goes to town to sell a load of pine and hickory in stove-wood lengths.

When Tom cuts his trees, he does it carefully. He takes out the trees which are mature and those which are badly bent. He thins out trees which stand too close together. This kind of cutting and thinning leaves room for young trees to grow.

In a way Tom treats his trees as if they were a crop. He plans to harvest this tree crop every few years. If he had more land, his tree crop would be an important part of his yearly earnings.

The cost of raising trees is not large. Trees grow fast in North Carolina. So with proper methods of cultivation and cutting, and with protection from fires, nature does the rest.

A number of the plants in the Carolina Mountains are used in making medicine. They find their way into chemical laboratories and drug stores all over the world. Three-fourths of all the drug plants collected in the United States are gathered by farm and mountain families in the country within a radius of one hundred miles of Asheville. And many farmers are now cultivating some of the drug plants as a regular crop.

Galax leaves and other evergreens are picked and shipped to city markets. Rhododendrons, azaleas, and other native plants are dug and sold. Nature's wilderness garden is a source of wealth to the people.

Beech trees grow tall and straight in Pisgah National Forest.

The leaf blade of the Venus Flytrap closes when anything enters.

The beauty of trees and plants is a part of the state's wealth. Travelers come from other states to see North Carolina's mountains covered with trees—evergreens lost in the new green of spring or dark among the brilliant leaves of fall. These visitors have heard of the way the highlands flame with azaleas, of the great rose-colored rhododendrons spilling across the open spaces of the mountains. They discover for themselves how dogwoods all over the state reach out from the shadows of the trees into the sunshine.

This beauty, which others must come here to enjoy, is wealth to North Carolinians, who have it at their doorsteps. It is a kind of wealth which cannot be deposited in a bank or spent for the week's groceries, a kind of wealth which we often overlook or take for granted. For many people it is a kind of wealth which they would not like to lose.

The silhouette of the state gives variety to North Caro-

The plant feeds on flies and other insects which are caught in these traps.

lina's plant life. As the land rises from coast to mountains, there are varying temperatures, soils, moisture, and drainage. And these different growing conditions produce many different kinds of plants. Some cannot live if they have too much water, and some die if they have too little. Some thrive in the woods or on the cold mountain tops, and some manage to survive when the weather is very hot and very dry.

In moist sandy bogs not far from Wilmington grows one of the wonders of the plant world—the Venus Flytrap, which is found only in certain parts of North and South Carolina and nowhere else in the world. This curious plant is of interest to botanists because it feeds on insects caught in traps on its leaf-blades.

Altogether there are over six thousand different kinds of plants in the state. Botanists who study the How, Why, and Where of plant life find a natural laboratory here in North Carolina.

Here Are Some Questions Calling for Further Study

1. How many things do you use daily which are made from wood?

2. What are the most common trees growing in your community? The pamphlet by Holmes, *Forest Trees of North Carolina*, will help you identify them.

3. What determines the kind of trees found in an area?

4. What do trees have to do with preventing erosion, controlling floods, supplying water, and protecting wildlife?

Here Are Some Interesting Things to Do

1. Let each member of the class choose one tree in your neighborhood to "live with" during the year. Notice how the tree changes from time to time. Study its leaves, limbs, bark, flowers, fruit, etc. Take or draw pictures of it from month to month.

2. See if you can start a movement for a School Forest in your community. Write for suggestions about this to the Forestry Division of the State Department of Conservation and Development in Raleigh, or to U. S. Forest Service in Washington, D. C.

3. Arrange an exhibit to show the wildflowers and trees of your neighborhood. Pin flowers and leaves to pieces of paper and label each one carefully.

4. Try your hand at making toys, gifts, and decorations with products from the out-of-doors near you. There are many things you might do.

5. Read and report on *Your Forests*, by Bruere, and *Trees of the South*, by Green.

8. *Wildlife*

WHEN forests stretched from the east coast into the western distance of this "fair and spacious Province," wild game was plentiful. The Indian, hunting and trapping for his simple needs, did little to disturb the balance of nature. All living creatures shared the wealth of the land alike—Red Men, animals, birds, and fish. Indians killed for their food, just as the others did.

This changed with the coming of the white man. His gun was deadly, and it made a lot of noise, frightening the animals it did not kill. His traps were sharp and strong. His ax cut into the forests of longleaf pine, cypress, gum, and oak.

As the white man advanced, wildlife retreated. Today, some of the animals which used to range North Carolina have disappeared. There are no elk hunting through the shadows for tender shoots of willows or birch. It is years since native beavers built their dams in North Carolina streams. The otter is becoming scarce.

North Carolina has brought elk and beavers from other states and has placed them in game refuges where they are safe from hunters. After a while, this natural resource may be partly restored.

Men on the coast no longer catch a glimpse of Carolina

White-tailed deer follow narrow paths through the underbrush in the mountains and in the low country along the coast.

paroquets. But perhaps the people who have apple trees growing in their yards are not sorry that these green and orange birds are extinct. When William Byrd helped to survey the boundary line between Virginia and North Carolina in 1728, he wrote, "Very often, in Autumn, when the Apples begin to ripen, they are visited with Numerous Flights of paraqueets, that bite all the Fruit to Pieces in a moment for the sake of the Kernels."

Migrating passenger pigeons used to fly over the mountains and the Piedmont in flocks so large they darkened the sky. They, too, have disappeared, not only from North Carolina but from the United States.

Although some kinds of wild animals have decreased or disappeared, North Carolina still has some of the best hunting in the country. And there are now state and federal laws which protect fish and birds, game and fur-bearing animals. Because of these laws, wildlife will survive and be a source of pleasure and wealth not only today but in the years to come.

In the mountains and in the low country of the coast, men hunt black bears and white-tailed deer. Muskrats, minks, weasels, skunks, and fox are trapped for their furs. On clear cool nights, when the moon rises late, men and dogs follow the trails of raccoons and opossums over the lonesome hills. Boys in every county go out with their guns and bring back squirrels or rabbits for a stew.

Quail and doves range over the whole state. Not long ago these birds were shot in such numbers that they were becoming scarce, but now the game laws protect them.

Quail eat potato weevils and other insect pests like wireworms, grasshoppers, squash-bugs, chinch bugs, and caterpillars. Doves and quail eat the seeds of many weeds —crab grass, sheep sorrel, bindweed, chickweed, and ragweed. Perhaps this diet of pests makes up for the ripened wheat on which the birds sometimes feast.

Turkeys, the largest game birds in this country, are fairly common in Carolina's eastern swamps and highland forests. And they appear in small numbers even in the central part of the state, along the river bottoms. Hunters must be skillful to get near enough to shoot turkeys, for they are always on their guard.

Tom Caswell stalks them sometimes, using a "turkey call" which his grandfather made from a gobbler's wingbone long ago. Off in the woods, a turkey hears the call and comes near to see this "other turkey." But he is a

Men hunt quail in every section of North Carolina.

cautious bird. If the call sounds strange, or if he catches a glimpse of Tom, he disappears, dodging behind trees and hiding in underbrush. Hunting turkeys is good sport, and a man is proud of his skill when he brings in one of these great game birds of the world.

Migratory wild fowl are plentiful. Large flocks of them fly south for the winter from their breeding grounds in the cold north and northwest. Many go no farther than North Carolina's coastal sounds, or lakes like Mattamuskeet.

In early autumn, thousands of geese arrive from Labrador, Manitoba, or northern Quebec. They honk over waters which are called Currituck, from "Coratank," the Indian name for wild geese. A little later, flocks of brant settle on the salt waters of Pamlico Sound to feed on eel grass. Snow geese turn some of the sandy flats and beaches of Currituck County's outer banks into "snow banks." Whistling swans winter in the shallow waters around Currituck Sound.

The beautiful wood or summer duck is here all year, and is joined in the autumn by others of its kind from the north. The mallard and the black duck "tip up" for their food in shallow waters throughout the state. There are teal, pin tails, and baldpates. On the open waters of Currituck Sound, redheads dive for their food. And canvasbacks float together on the water in flocks like large rafts. Altogether, North Carolina has more than thirty different kinds of ducks.

So the sandy reefs which make an island wall along Carolina's coast once again play a part in the state's wealth. In the fresh-water and salt-water sounds, among the marshes and along the protected beaches, wild fowl find what they are looking for, food and sanctuary. Some

Whistling swans winter in the shallow waters around Currituck Sound.

of the birds eat shellfish, sea worms, mollusks, and small fish. Some go ashore for acorns and grain. Most of them feed on the plant life growing in and around the water—bushy pondweed, tape-grass or "wild celery," eel grass, and other grasses.

At times, something disturbs the abundant plant life. Disease almost destroyed the eel grass which was once plentiful in Pamlico Sound. In 1933, when the ocean broke across sand reefs into Currituck Sound, salt water killed the "wild celery" and other fresh-water plants. When food is scarce, wild fowl are scarce too.

Hunters have done more than the lack of food to decrease the number of birds along the coast. Before the game laws of 1918 there was no real sanctuary here for the huge flocks migrating from the north. Year after year men shot quantities of ducks and geese and shipped them to market.

Those men were making a living by using one of the sources of natural wealth, but they were doing it wastefully. They were destroying the source of their income. The American brant, the redhead, and the canvasback were in danger of becoming extinct.

Then federal and state game laws were passed. Men were no longer allowed to go out in boats at night, blind the ducks with bright lights, and kill them before they were able to rise from the water. For several years hunters were not allowed to shoot any of the species which had been so popular. In the 1930's, National Wildlife Refuges were established in the state, and migrating birds now find resting spots there—on Mattamuskeet Lake, the Swanquarter Refuge, and the Pea Island Refuge. North Carolina's resources in wildfowl have been saved.

North Carolina's wildlife offers another kind of sport to many people. Instead of hunting with guns, they hunt with field glasses, cameras, and notebooks. They study the habits of birds and other animals, take pictures of them, often make notes of what they see. This kind of hunting may be as much sport as hunting with guns.

Wildlife is a source of wealth all over the state. Men know they must preserve this resource just as much as the resources of soil and trees and water and fish.

For many people, "hu ing" with field glas cameras, and notebook more fun than hunt with guns.

Do You Know

1. Which are the most common birds in the state?

2. What animals and birds are hunted for sport? For food? For fur?

3. Why some animals and birds are spoken of as migratory?

4. Why North Carolina has such a variety of wildlife?

5. What is being done to protect wildlife by such groups as North Carolina Bird Club, The Audubon Society, Fish and Wild Life Service of the United States Government, the State Wildlife Resources Commission, and 4-H Wild Life Conference?

6. What other groups help protect wildlife?

Have You Read or Seen

1. Green, *Birds of the South?*

2. Hammonds and Woods, *Today's Agriculture*, Chapters 18–20?

3. Pearson, Brimley, and Brimley, *Birds of North Carolina?*

4. *The Chat*—a magazine published by the North Carolina Bird Club?

5. *Wildlife in North Carolina*—a magazine published quarterly by the Wildlife Resources Commission?

6. Parker, *Our Friendly Neighbors?* This is a beautifully illustrated book of poems about birds and animals found in town and country in North Carolina.

9. *Beneath the Soil*

A PART of North Carolina's natural wealth, its mineral wealth, lies beneath the soil. The state has a great variety of minerals.

What makes our minerals so varied? The answer may be found in the geologic history of North Carolina. Minerals are varied because there are different kinds of rocks in the earth under our feet and because these rocks have changed and broken up in various ways during the past centuries.

This variety is itself a large part of North Carolina's mineral wealth. Altogether there are over three hundred species and sub-species of minerals here. The state is a "sample case" of the world's riches. It is a geologist's natural laboratory.

Buried in the earth are copper, manganese, chromium, iron, nickel, zinc, tin, and many other metals. There are even places where men may dig out the gold and silver that Ralph Lane's men wanted and never found.

But only a few of North Carolina's metallic minerals are a source of wealth. Only a few can be produced in quantity and sold at a profit. Most of our metals lie untouched. Why? There are other places where these minerals are more plentiful, and where they can be mined more easily and sold more cheaply. The metals in North Carolina are only "sample-case" wealth.

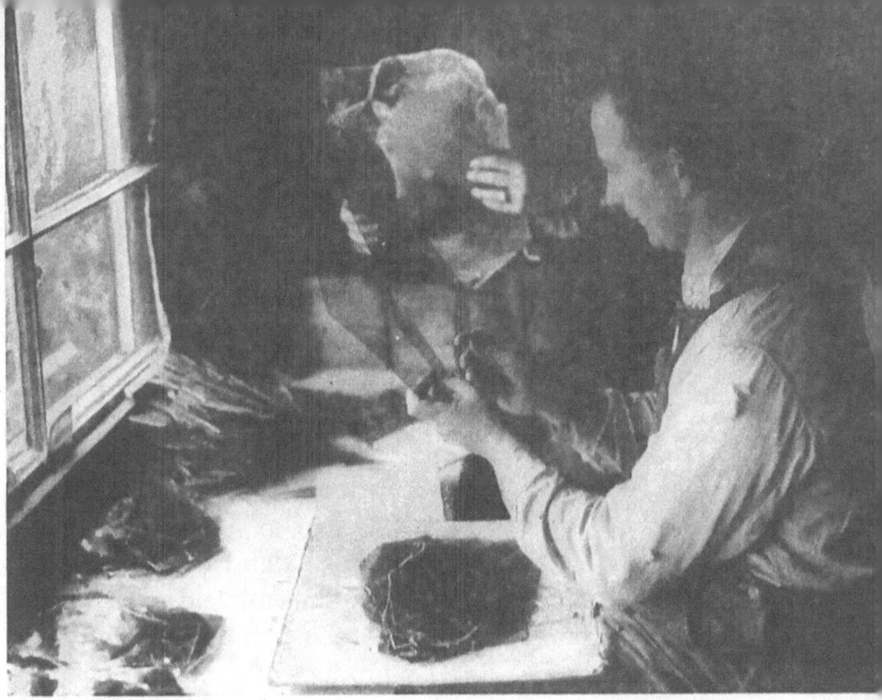

Blocks of mica are peeled by hand in thin sheets. The larger the sheet, the more the mica is worth.

Besides metals, many different kinds of precious stones have been mined in the state—the sapphire, ruby, garnet, beryl (aquamarine, yellow beryl, and emerald), the hiddenite or lithia emerald, rock crystal, amethyst, smoky quartz, even a few diamonds. Some of these stones are clear and without flaws. They are true gems.

But so far, only the beryls and the garnets have paid men for mining and marketing them. And only the commoner garnets are mined—those which are used in grinding and polishing hard metals like steel. North Carolina's important minerals are not the precious stones.

Then where is our mineral wealth? The state's mineral wealth is found in simple non-metallic minerals—the clays and the rocks.

Many parts of the state have clays suitable for making

brick, tile, and pottery. And there are special clays in various sections.

Feldspar is a clay used in making glass, pottery, and earthenware. It is found chiefly in the country around Spruce Pine, in Avery, Mitchell, and Yancey counties. The amount of feldspar produced there is large.

Kaolin is found in large quantities in Avery, Mitchell, and Yancey counties. North Carolina kaolin, which is of a very high quality, is used in making china dinnerware. Some specialists say that it is just as good as the clays used in the fine china for which England is famous.

A neighboring mineral of kaolin and feldspar is mica. All three are found in the same part of the state. And mica is sometimes mined as a by-product of kaolin.

Mica is transparent. It was once fastened over small openings in the doors of coal- or wood-burning stoves and heaters, as "windows" through which people could look at the fire. There are still stoves in the country with these mica windows.

But most of the mica of today has more modern uses. It has a place in electrical machinery; in paints, wall papers, and lubricating oils. It is used in roofing, and in materials for fireproofing and insulating.

Men dig mica in mines beneath the ground, and bring it out to the surface in blocks. Then other workers peel the blocks into thin sheets. The larger the sheet, the more the mica is worth.

This is the way people have been mining mica for a long time in the state. But a great deal of the mineral is wasted in this way. Some is lost in the mines, and some is lost during the process of peeling because the mica chips and breaks. If people could find better methods of mining and handling mica, it would produce a great deal more wealth.

Vermiculite is another North Carolina mineral. It is a hydrated form of mica which is highly resistant to fire. When vermiculite is exposed to great heat, it flakes and becomes very light. It can be used in place of some of the sand and crushed rock in the concrete of buildings. Vermiculite in the concrete helps to make buildings more fireproof.

North Carolina has large deposits of olivine, a mineral which may be used in making magnesium. Magnesium is a modern, lightweight metal which is becoming more important in manufacturing various articles. It often takes the place of another lightweight metal—aluminum.

There is spodumene in North Carolina—a mineral which is used in ceramics and metallurgy. And the state also produces talc, soapstone, pyrophyllite, a little asbestos, and small quantities of some of the other minerals in the sample case.

Granite from North Carolina quarries is used for buildings in many states.

With these the list of North Carolina's minerals is almost finished. Almost, but not quite. There is one group left. And while this last group may be the most commonplace of all, it is probably also the most valuable. In this group are the minerals used in buildings and in roads.

North Carolina has deposits of granite and marble and other kinds of rocks and stones out of which we make some of our most beautiful buildings. We have plenty of common rock to be crushed for our roads. This rock is abundant all over the western half of the state. So whenever concrete roads are built, the necessary rock is generally not too far away.

In May, 1947, a mineral research laboratory, located at Asheville and operated by State College, was officially put in operation.

Perhaps some day a use which men have not yet dreamed of may increase the value of North Carolina's mineral resources. Until then, the granite and the stones for buildings and the clays for bricks, tile, pottery, and china probably produce more wealth than any of the other minerals. And the rock which is crushed for our roads and concrete buildings is perhaps the most valuable mineral of all, not in money produced but in money saved. This common mineral has saved millions of dollars for the people of the state.

Be Sure to Learn All of These

1. The difference between metallic and non-metallic minerals.

2. The reasons why North Carolina is called "Nature's Sample Case."

3. The chief uses of each of these: granite, feldspar, kaolin, mica, vermiculite, talc, soapstone, pyrophyllite, and asbestos.

4. The importance to the state of its building stones.

5. The native product from which our state capitol is made.

Be Sure to Do Some of These

1. Read what your science books say about minerals.

2. Unless your school already has them, appoint somebody to write to the State Museum in Raleigh for copies of all their mimeographed circulars, especially those dealing with minerals and gems.

3. Divide the class into committees and let each committee study and report about one of the state's leading minerals.

4. Collect, identify, and label samples of as many minerals as you can. If you need help, consult such a book as *Minerals, Their Identification, Uses, and How to Collect Them,* by Zim and Cooper. To which State Department would you send your specimens in rocks or stones to be identified for your permanent school collection?

5. Visit a pottery, rock quarry, or mine, if there is one near by. Try your hand at making pottery or other articles from North Carolina clay.

To mine kaolin and feldspar, men often dig large open pits in the ground.

10. *Conserving Our Natural Wealth*

THIS, then, is the natural wealth which is one of the great resources of North Carolina today.

We have soils with varied richness and fertility. We have long growing seasons for many kinds of crops, with moderate temperatures and frequent rains. We have lakes, streams, and rivers, coastal sounds, and the ocean itself. There is water to be used as highways, and water to make power for turning industry's countless wheels.

We have beaches and mountains, sunshine, scenery, and sports to make the state a tourists' vacation land from the Atlantic to the Smokies. Winters are so mild that much of the time people can work or play out-of-doors. There are fish, wild animals, and birds for food and for sport. There are trees and plants and minerals.

This is the wealth which men have used for years—often wastefully because for years there seemed to be no limit to the resources.

How have we wasted our natural wealth? First by the way we farmed the land.

Fields were planted with the same kind of crop season after season, until the humus and plant food were used up and the soil was no longer able to grow a crop. Then land was cleared for new farms. The farmer and his family

The farmer and his family moved away, and the worn-out land was left to the rains.

Forests were cut from steep hills and mountains, and no young trees were left standing to hold the soil in place.

Flood waters from the bare hillsides killed the crops growing in the low bottom lands.

moved away. And the old fields were left to the winds and the rains.

If the soil did not wash away too fast, crabgrass covered the old field during the first year, followed perhaps by ragweed in the second year and broom-sedge and brambles in the third. Four or five years after the field was abandoned, pines sprang up, held their own for seventy-five or a hundred years, and then were driven out by oaks and hickories.

When a field like this is left undisturbed, beeches and maples may take the place of the oaks and hickories. Finally the richness of the soil will be restored, and a forest will once more cover the land just as it did long ago.

But this will take two or three hundred years more. It is perhaps four hundred years in all before nature, unaided, can repair the damage done by men.

How else did we waste our natural wealth? By the way we cut down our forests. For the sake of the lumber and the wood pulp, trees were often cut from steep hills and mountains. No young trees were left standing to hold the soil. The land sloped too sharply for cultivated crops, and nothing was planted to take the place of the forests. Before very long the topsoil washed away. Erosion cut the hills and mountains into gullies that deepened and widened with every rain. The land was wasted.

The soil was lost. But the loss in natural wealth did not stop there. The soil was carried by rain into the streams, where it damaged other resources. Water which had been clear became so muddy that it was sometimes unfit for human or industrial use. Fish became less plentiful. The

---

← HOW WE HAVE WASTED SOME OF OUR NATURAL WEALTH

soil and sands in the rivers gradually settled to the bottom, filling in the river beds.

When it rained, these smaller river beds were not able to carry away the larger amount of water that came rushing down the cut-over slopes where there were no longer any trees or carpets of leaves to hold the moisture. Then the rivers overflowed their banks. The flood water often killed the crops growing in the low bottom lands. It ruined the soil there by leaving on top of it the sand and gravel from the mountains and the hills.

Thus the wealth of the land has been wasted and lost.

Today men are conserving this wealth. Many are farming so that they will not use up the rich fertility of the soil. They do not always plant the same crop on the same land. After a field has been growing cotton for a year or two, Tom Caswell changes to some other crop for a year. Perhaps it is corn, with cowpeas or soybeans between the rows. Perhaps it is small grain or some other crop that does not have to be cultivated.

This kind of change is called rotation of crops. Crop rotation helps to control weeds, disease, and insect pests. It also helps to keep the soil fertile.

Sometimes the land which grows cotton is not suited to corn, and the land which grows corn is not suited to cotton. Then the soil may be improved during the mild winter months. In the fall Tom plants vetch, Austrian winter peas, crimson clover, or one of the other good cover crops. The winter crop keeps the rains from washing away the soil.

Before spring, Tom plows the crop under. This green manure, as it is called, improves the soil by adding plant food and humus. In the spring, corn and cotton may be planted where they were the year before. And in the fall, the yield is better because the cover crop made the soil better.

Men are saving the wealth of the land by controlling erosion. For years North Carolina farmers planted corn and cotton and other row crops on hillsides which were so steep they should never have been broken by a plow. It did not take long for the topsoil to disappear down the slopes into the streams and rivers.

These slopes are now being planted with crops that do not have to be cultivated—apple orchards, grapes, pastures for cattle, or trees that restore the forest wealth. They all keep the soil where it belongs.

Even the fields which are suitable for farming are often hilly and rolling. To hold the soil in place, farmers are beginning to plow and plant by methods which control the flow of water. They plow their furrows around the slopes, not up and down them.

Sometimes they plant crops which cover the land and do not have to be cultivated. They use strips of these cover crops between the cultivated crops, and the strips slow up the flow of surface water.

Wherever the water has made small channels to lower levels, the farmers plant permanent grasses which keep the channels from deepening into gullies. When gullies have already formed, the farmers build dams of rocks, brushwood, and saplings. These dams let the water through, but they keep it from running away with the soil.

Men are beginning to repair some of the damage which has been done to North Carolina forests and woodlands. There are thousands of acres in the state which were abandoned long ago. They are now so worn away by erosion that they are worthless for farm crops. Tree seedlings are being planted on these wasted lands, and the soil and forests will be restored.

Farmers will not need to clear new land if they do three things—(1) plant crops best suited to the soils in their old fields, (2) rotate these crops, and (3) keep the soil fertile. Fewer acres have to be abandoned, and more woodlands are left uncleared.

Trees may be cut for lumber and pulpwood in a way that makes sure there will be new trees growing on the land to hold the soil and produce another tree crop some time in the future. Forests can be protected from the danger of fires, which in the past have wasted trees, soil, birds, animals, fish, and streams.

Nature keeps a balance among all living things. When this natural balance is thoughtlessly upset, the result may be unexpected—perhaps good for one year or for ten years, but bad fifty years later. So we study the past and the present. We learn what was done years ago to make things the way they are today. Then we know better what to do today so that our natural resources will be saved for the people of tomorrow.

When men use the land, they must not abuse it. They no longer farm just for the crop of this year or next year. They use the soil so that it will yield a crop year after year. They conserve, or save, the wealth of the land.

Men realize that in a way the land is never theirs, even though they may own the same acres which their fathers and grandfathers owned before them. All men simply hold the land in trust for the generations to come.

HOW WE ARE CONSERVING SOME OF OUR NATURAL WEALTH →

Land that slopes sharply is now being planted with crops that need little or no cultivating. The orchard on this hillside keeps the soil where it belongs.

Farmers plow their furrows around the slope, instead of up and down. A crop like tobacco grows well in rows which follow the curve of the land.

Crops on sloping fields may be planted in wide strips. Here they are strips of corn and of hay. The hay is a permanent crop that covers the land and makes the water run slowly down the hill.

Tree seedlings planted on our wasted land will stop erosion and cover the land with forests once more.

Here Are Some Ways to Learn More About Conservation

1. Compare the color of soil from the woods with soil from a field in which crops have been planted for several years. Which is darker? What makes the difference? What seems to be one effect on soil of growing crops in it year after year? Also test the two types of soil to see which holds water better. Put equal amounts of each in two tin cans which have holes in the bottom. Place cans in pans. Pour equal amounts of water into the cans. Which loses more of the water?

2. Examine the roots of alfalfa, peas, red clover, or some other legume to see if you can find tiny enlarged places called nodules. What do these nodules do? Why is the planting of legumes an important step in conserving the soil? Notice how the roots of some plants are matted together. How do plants like these help to hold the soil?

3. Look for signs of erosion on your school grounds. What caused this erosion? What can be done by your class to check it?

4. Make a trip to a near-by farm where approved methods of conservation are being carried out. Your agriculture teacher, your county agent, or the Soil Conservation representative will be glad to help you.

5. Make a picture graph (or pictograph) to show the amount of damage done by forest fires in North Carolina over a period of years. Show in another graph the chief causes of fires. How can you go about getting the information for this graph?

6. If you have a movie projector in your school, you will want to see some of the many good films about conservation, such as *A Heritage We Guard*, or *The Conservation of Natural Resources*. See whether your county farm agent or local Soil Conservation official can secure them for you. If not, write to the Bureau of Audio-

Visual Aids, University of North Carolina, Chapel Hill, for a catalogue listing these and other films.

7. Divide your class into committees to study and report on the agencies which are carrying on conservation work in the state—and don't forget private agencies.

Here Is an Outline to Use in Making a Summary of What You Have Learned

I. How North Carolina's natural wealth has been wasted.

II. What must be done to conserve the natural wealth.

III. What agencies are working toward conservation.

IV. What my school can do and how I can help.

Here Are Some Questions to Discuss

1. What is meant by saying that soil is "an indispensable resource"?

2. In what way is water an "inexhaustible resource"? In what way is it not?

3. In what way are forests "a renewable resource"?

4. Why are minerals classed as "exhaustible resources"?

5. In what ways is wildlife a "manageable resource"?

6. In what way do terraces "make running water walk"?

7. What is meant by nature's balance? What examples of it can you give? In what way has man upset this balance? To help you answer these questions, read *Balance in Nature*, by Bertha Morris Parker.

Here Is a Book to Read and Report on

Conservation of the Nation's Resources, by Flynn and Perkins.

11. *The People*

RALPH LANE'S men left the natural wealth of the land untouched. And after the Lost Colony disappeared, the wealth lay undisturbed by white men for years.

Later—no one knows exactly when—settlers came to live in the "fair and spacious Province" which is now North Carolina. But they came not by sea, for the islands still stood like a wall along the coast. They came by land, from the north.

These men were not hunting gold, or silver, or precious stones. They were planters. They had heard that land to the south was plentiful, cheap, and fertile, the winters so mild that horses and cattle could find their own food without help.

Who were these men? First, Englishmen came from Virginia to the land above the Albemarle and made it their own. They brought something of England with them—language, religion, a love of liberty and the will to defend it even if they had to fight for it.

As their numbers grew and years passed, they spread along the bays of the coast and westward across the plain. And as they cleared the rich lands of the plain, they brought in Negro slaves from Africa to help farm their huge fields.

Swiss and German families settled between the Neuse and Trent rivers. Farther to the south, Highlanders from Scotland settled on the upper Cape Fear. These Scots were strong men, fearless and wild, used to fighting among themselves and with their neighbors of the lowlands.

To the west of the Cape Fear settlement came Scotch Lowlanders who had been living in Ireland until they were forced to find homes somewhere else. These Scotch-Irish had gone first to Pennsylvania. Then many of them moved on to the rolling country which we call the Piedmont. Before the Revolution, they had spread through the hills and valleys and along the river banks of western North Carolina.

About the time that the Scotch-Irish were making their way southward along the "Wagon road" to the valleys of the Catawba, Eno, and Yadkin rivers, Germans were following the same course marked out by the foothills of the Alleghanies. Some of them came to hunt and trap. Some came because the land was cheap. Some came to find freedom of religion. Instead of claiming large tracts of land and living far apart, many of them settled close together in one community, and built churches and schools as well as homes.

The German farms were smaller and better managed than those of the English. The German farmers, working for themselves, were more careful and more skillful than the Negro slaves working for the English planters. Many of the Germans spoke no English, and they lived on remote frontiers far from the coast where government was centered. They took little part in politics. They were willing to leave such matters to the English and the Scotch.

These were the settlers destined to use the wealth of the land—the Englishmen, Scots, Scotch-Irish, and Germans;

the French, Swiss, Dutch, and Welsh in smaller numbers.

The first colonists built their homes and cleared their fields close to the rivers and the sounds. Travel was easier by water than by land because there were no roads and forests stood thick and tangled.

Then, as space along the water became scarce, pioneers pushed inland to the unclaimed wilderness across the Coastal Plain. Later settlers found their way to the Piedmont, and by 1760 families were scattered as far west as the Blue Ridge.

In 1790, when the first Census was taken, there were almost 400,000 people in the state, not counting the Indians —about 300,000 whites and about 100,000 Negroes. After 1800 there was very little immigration. The population increased naturally as children grew up, married, and had children of their own. Families were large, and gradually the population spread over the state.

The early settlers had come from different countries. They worshipped God in different ways. They had different languages and traditions, different songs, different food, different kinds of dress.

At first each group kept to itself, and looked on the others as strange. Then families moved from one settlement to another, perhaps to new land where the nearest neighbors were these "different" people. They learned to like these neighbors, they intermarried with them, and after a while there were no longer English, Scots, Scotch-Irish, or German, but a blending of nationalities, families, and inheritance. They were all North Carolinians.

In 1940 there were more than three and a half million people living in North Carolina. There were 2,567,635 white people, 981,298 Negroes, 22,546 Indians, and 144 people of other races.

These people are the state's greatest resource. Here are people to use the wealth of the land, people to grow crops and to work in factories, people who produce for themselves and for others, people who need some of the things which others produce. Here is our great human wealth—hands, thousands and hundreds of thousands of hands.

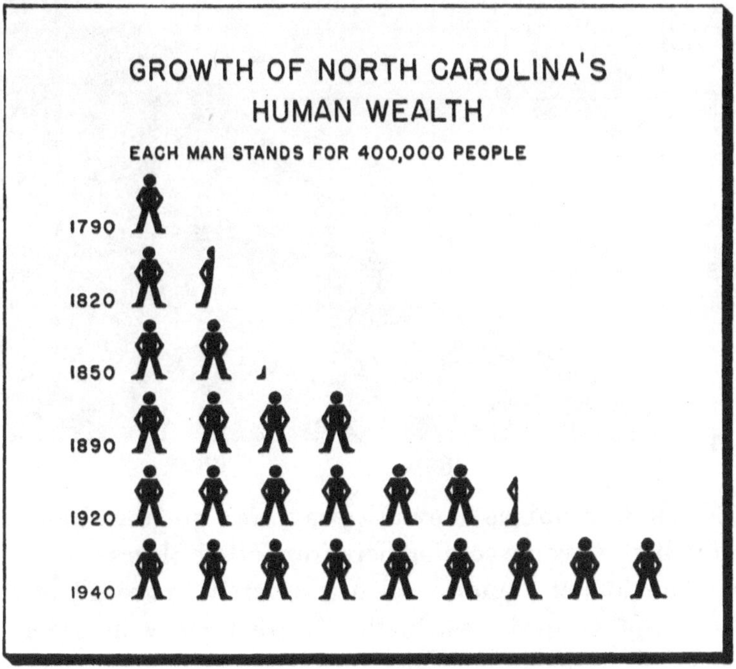

Most North Carolinians today are native-born: descendants of the men and women who came here before 1800. Only a few of the people living here, about one hundred out of every thousand, were born in other states. Only three in every thousand are foreign-born.

Instead of having people move here from other states and countries, North Carolina in the past lost some of her own population to other states. For years the number of

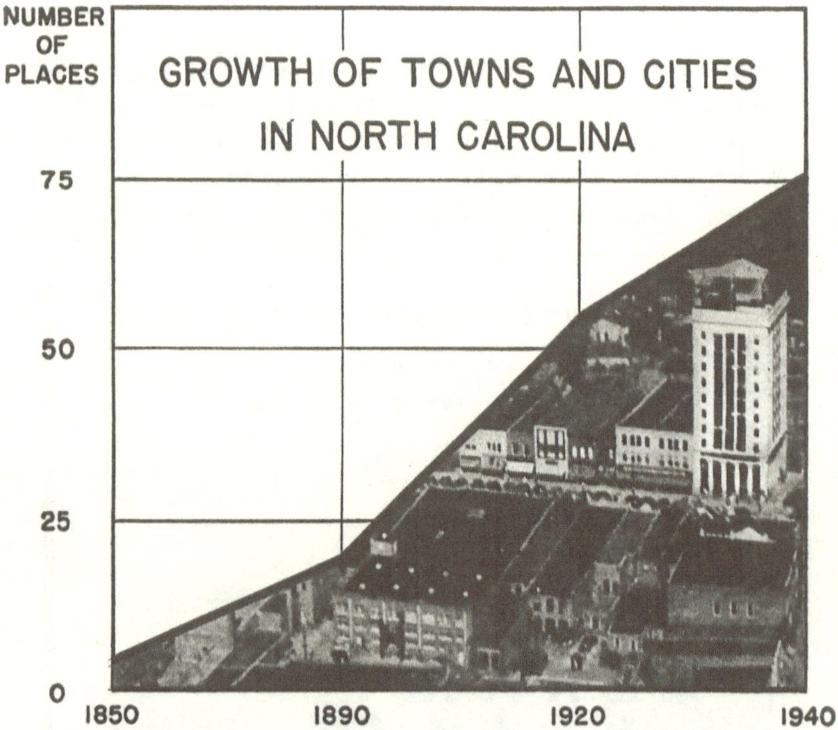

North Carolinians moving away was greater than the number of people coming here from other places.

This loss of human wealth was partly the result of the waste of natural wealth. People went away in order to make a better living.

North Carolina has always been a rural state, with many of the people making their living on the land. Their living became more and more difficult after men had used the wealth of the land wastefully, and the fertility of the soil had been lost. Other states and other countries often raised better and cheaper crops.

It was hard for farmers to take care of their large families, to feed the children and buy clothes for them. It was even harder to send them to school and to get doctors and medicine when they were needed.

People left their homes. They moved to neighboring states and distant states, to the West and to the North, wherever they thought they might have a chance to make a better living. The state was losing human wealth.

Then the factories in North Carolina, expanding and increasing, had work for men and women to do—simple work for hands which had only simple skills. Here was opportunity for people who were looking for jobs, people who had to make a living, people who wanted to better themselves.

Towns and villages grew larger as people looking for work moved in from the country. Cities appeared here and there, especially in the Piedmont, where manufacturing made jobs for many men and women. North Carolina was no longer entirely a rural state.

People from other states began to come here. And after 1900, North Carolinians who had decided to leave their farms and mountain cabins often went no farther than the nearest factory town.

For a while the new way of life seemed good to a country family. There was regular pay every week, perhaps more money than they had been used to seeing in a month. Sometimes there was a comfortable house, occasionally with electricity and running water. The noisy, brightly-lighted streets were crowded with people who had known the lonely silences back home. There was always something pretty in the dime-store window, something good to eat at the corner grocery. Money did not last very long.

And then times changed. In the depression years of the 1930's, factories were making fewer things, because many men and women had no money to spend. People looking

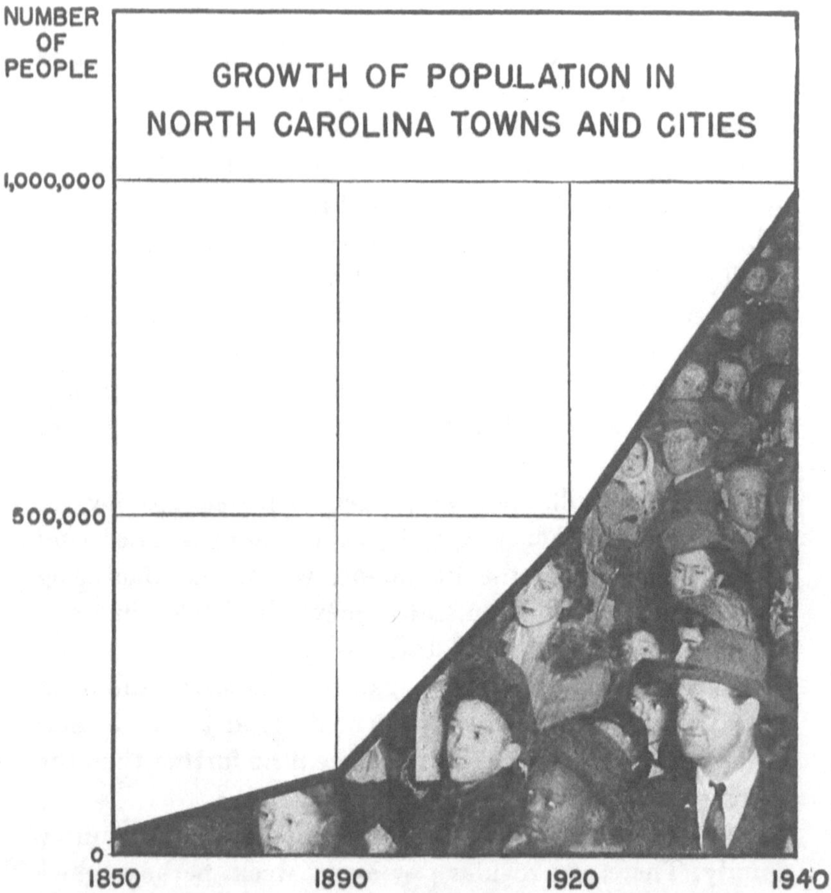

for work could not always find it. There were more workers than there were jobs. Pay was low. Sometimes new machines were hauled in and set up on the factory floor. One machine, tended by one man, could do the work which many men had been doing.

The drift of North Carolina's population changed. Some of the people who had moved from farms to factories went back to the farms again. Perhaps they could scratch a living on the land, and would not starve. In 1932, there were more people going back to the farms than there were people leaving them.

Many of the factory workers had nothing to do in town, but they had no way of going back to the country and nothing to go back to. They had to have help in order to live. The people who had never left the country stayed there, although it was harder and harder to make a living.

By 1940 there was work in the factories again, and people started moving back to towns and cities. After the United States entered the Second World War, even more workers were needed. Many men who were not in the armed forces left their farms to make guns and ships and other war supplies. And many women did the same thing. Towns and cities in North Carolina grew larger and larger.

Was it any easier for farmers to make a living, after people began moving away from the land? During the 1920's life in the country should have been both easier and better. There were fewer people depending on the farm income for a living. And automobiles and roads, telephones and radios and electricity were all making it possible for people to live more comfortably on farms, to be a part of what was going on in the world.

There were more people in towns and cities, and they had to buy their food. There were fewer people to grow this food. But North Carolina farmers never took full advantage of this market. Fruits and vegetables which they might have grown and sold were brought here from states as far away as Texas, California, Oregon, and Washington.

In 1944 the average total income per farm in North Carolina was higher than it had ever been before: $2,993. Yet in thirty-seven other states the average farm income was higher.

In August of 1946 the average wage paid to the men and women working in manufacturing industries in North Carolina was $34.20 a week. The weekly average in manufacturing industries in the United States was $45.10.

What is the reason for this difference? Why are the average incomes of North Carolina farmers and industrial workers so much lower than the average for the nation? The answer lies partly in the skills and training of the people, in their machines and equipment, in their methods of using the wealth of the land.

The records show that this difference in income is growing less as our training and our skills become greater. But there is still a great deal to be done if we are to make the best possible use of our resources.

You'll Want to Learn More About North Carolinians

1. Read and take notes on these references:
   a. Arnett, *The Story of North Carolina*, pp. 63–64; Chapters 16, 17, 18.
   b. *North Carolina: A Guide to the Old North State*, pp. 3–7, 25–30, 51–57.
   c. Newsome and Lefler, *The Growth of North Carolina*, pp. 30–35, 72–80.

2. Try to find who the early settlers in your community were. Make a booklet containing this information.

3. Show in a graph the racial distribution of the state's population according to the most recent census. Make a similar graph showing this information for your county.

4. Think about and discuss these questions:
   a. In what part of the state do most of the people live?
   b. Why do so many live there?

c. Why does North Carolina have so few foreign-born citizens?

You'll Enjoy

1. Learning songs and dances typical of the different racial and national groups who settled in the state.

2. Writing some poems about North Carolinians.

3. Reading and reporting on these books about North Carolinians—some real and some imaginary:

a. Allee, *Road to Carolina,* and *Susanna and Tristam.* Two novels of the Civil War period.

b. Brawley, *Negro Builders and Heroes.*

c. Connor, *Makers of North Carolina History.*

d. Fitzgerald, Mary Newman, *The Cherokee and His Smoky Mountain Legends.*

e. Gray, *Meggy McIntosh.* A story with a Cape Fear setting in the time of Flora MacDonald.

f. Hamlin, *Ninety Bits.* Short biographies of ninety North Carolinians in different fields of work.

g. Hendricks, *Bundle of Troubles and Other Tar Heel Tales.* Folk tales from different sections of the state.

h. Key, *With Daniel Boone on the Caroliny Trail.* A story about famous leaders with the Yadkin Valley as its setting.

i. Knox, *Gray Caps.* A novel about people in Raleigh during the Civil War.

j. Lenski, *Blue Ridge Billy.* A lively story of a boy in the North Carolina mountains.

k. Means, *Shuttered Windows.* A novel about a Negro girl who came from the North to live on an island off the coast of Carolina.

l. Worth, *The Middle Button.* A story of a girl in the Cape Fear region who wanted to be a doctor.

m. Worth, *They Loved to Laugh.* A novel about a Quaker family of Guilford County.

12. *Technological Wealth*

NATURAL resources are worth nothing unless men are able to use them.

North Carolina's settlers knew how to convert some of the wealth of the wilderness to their needs. Theirs was a "handmade" world. They cleared fields with hand-forged axes, the hickory handles shaped by hand with hand-forged knives. They built houses of hand-hewn logs. They turned their furrows with handmade plows.

They produced by hand all the things necessary for living. They made their own chairs, stools, benches, tables, and other furniture. Their clothing was sewed by hand of fabric they had woven from thread they had spun from cotton they had grown. They used wool they had clipped off sheep they had raised. They made shoes from leather they had prepared by tanning the hides of animals. They shaped pottery plates, churns, crocks, bowls, and other utensils from clays they had dug from the banks of streams. They hooked or braided or wove their rugs from scraps of material. They made baskets; brooms; iron pots, kettles, and skillets; tools; all the other implements for field and household.

With trained minds and skillful hands we are increasing our technological wealth.

Members of a family were often able to make everything for themselves. Well-to-do settlers had white servants and Negro slaves who were trained to make whatever the families used. Sometimes a person with special skill spent all his time making articles for other people. He was not always paid for his labor in money, because actual cash was very scarce. Many people paid him with things they had made.

Whoever did the work, the necessities of life were the result of hand labor. The skill with which settlers used natural resources was the technological wealth of their time.

This technological wealth increased through the years as men developed new skills, thought of new uses for raw materials or of new methods to produce what they needed. The word "manufacture" lost its original meaning, "make by hand." Machines were invented to do the work which hands had done. Steam produced power for the machines. Technology was changing the pattern of life.

Technological wealth becomes greater when we understand the machines we work with and find ways to do our work better and more easily.

How much of this technological wealth is there in North Carolina today? How skilled are the workers? How well trained are the people? How much do we know? And how much can we do?

To be able to answer these questions, it is necessary to ask others.

How many industrial laboratories are there in the state, and scientists studying to find newer and better ways of doing things?

What kinds of factories are there in the state? How many machines and tools are there, and how much of the work do they do? Are the people highly skilled? Do they have special training? Do they know how to make machines work for them?

The answers to these questions show that North Carolina does not rank high in technological wealth.

This machine for carding cotton is the product of technological wealth.

Industrial engineers and technicians are few. Many of the factories in the state carry on only simple manufacturing processes. The complicated work is done in other states. Much of the labor is still hand labor, with simple tools. And even these simple tools are not plentiful.

The people have skills, but they are generally the old skills, handed down from father to son. And many men today are not as skilled as their grandfathers were. Machines now make some of the things and do some of the jobs which used to be done at home. And men have developed few new skills to take the place of the old ones which are no longer needed.

North Carolina today does not have the technological wealth to make full use of its natural resources. Because of this, the natural wealth and the human wealth of the state have been wasted.

But our technological wealth is growing. With better education and more training, people will know more and can do more. When North Carolina has more skilled workers, our natural resources can be put to better use.

Can You Do These Things?

1. Define these terms: technological wealth, technician, laboratory, pattern of living, complicated process.

2. Explain what is meant by saying, "North Carolina today does not have the technological wealth to make full use of its natural resources."

3. Give examples of how the state's natural wealth and human wealth are not being fully used because of the lack of technological wealth.

4. Make a list of the things being done by your school to increase the technological wealth of North Carolina. Also list things being done by other schools or organizations near you.

13. *Capital Wealth*

BEFORE North Carolina's technological wealth can increase, there will have to be more capital wealth in the state.

More capital wealth would make many things possible. Industrial chemists and scientists could carry on the experiments and research which find better methods of production. They could invent better tools and machines, suggest new uses for raw materials.

With capital wealth men build factories and pay the men and women who work in them. Capital wealth produces railroads and highways and ships and trucks and machinery and tools—all the aids of living and of making a living.

Capital wealth is not just money in the bank. It is anything of value. It may be land, buildings, farms, work animals, tractors and other equipment, automobiles, trucks, machinery, or shops. It may be stocks and bonds, which represent something valuable. It is property. A person who owns capital wealth does not have to sell it in order to use this wealth. Because he owns wealth, he can borrow wealth.

Capital wealth is present in all kinds of business, large or small.

Some of this farmer's capital wealth is walking along behind him at the end of four ropes.

When Bill Caswell wants to earn some money, what does he do? He looks around, and finds out that no one has been delivering the morning newspapers to the people living along his street. So Bill goes from house to house, offering to deliver a newspaper, and soon he has a list of regular customers.

He does not need any actual cash to carry on this business, because he does not have to buy anything or pay anyone. He uses other capital wealth, but he does not use any money.

What kind of capital wealth would a boy have? Bill's is the bicycle on which he generally makes his daily deliveries. If he walks, and uses a wagon for carrying the papers, the wagon is capital wealth. And he uses his free time. Free time in which to work means capital wealth to Bill.

Bill's paper route itself is capital wealth. It is valuable because it is an established source of money. He can sell his paper route for cash, just as a man may sell his farm or his grocery store, his factory or his law business.

Bill's sister, Sally, is a member of a Scout troop which is trying to earn money. It is hot summer weather, and the girls decide to set up a booth where they will sell cold drinks.

The man who owns the vacant lot across from the post office lets them build their booth there. They pay no rent. They are using capital wealth—property. It is valuable

This machine in one of North Carolina's great paper mills stands for a large amount of capital wealth.

because of its location in the center of town, where many people pass by every day.

The girls draw money out of the troop's bank account to pay for lemons, sugar, ice, paper cups, and the other things they need. They are investing this money in their business. When they do this, they are using some of their capital wealth. They keep a record of their expenses, and they charge more for a cup of lemonade than the cost of making it. They return to their bank account the amount they withdrew to pay for their supplies. The money which they have left is their profit. They are using capital wealth to create more capital wealth.

When Henry Caswell decides to build a factory, he needs capital wealth. He has a sum of money which he will invest. He also owns something valuable—perhaps another factory, or an office building, or the land on which he plans to build, or some other land. This is capital wealth which makes it possible for him to borrow money.

But before anyone lends any money to Henry Caswell, he has to promise to pay something every year for the use of it. This is called interest. And he has to promise that within a certain time he will pay back the amount he borrowed, which is called the principal. He pledges his property that he will do this. If he does not pay it back, he may lose whatever he has pledged.

Henry Caswell needs capital to finance a new factory. He has to pay the men who design it, the men who build it. He has to buy the bricks, steel, and cement which go into its construction, the manufacturing equipment and the raw materials. Men and women have to be paid for their work.

These expenses are only a part of the cost of the manufactured article. Buildings and machines have to be re-

One young Tar Heel has found a simple way to change natural wealth into capital wealth. She sells grasshoppers, worms, and other live bait to fishermen. Here she is counting spring lizards into a jar for a customer.

paired. When machines wear out, they must be replaced. Sometimes a new invention makes expensive machinery useless and out of date. Insurance and taxes on the land and the factory must be paid. And there is also the interest on the money which Henry Caswell borrowed to begin with. And he must have the money to pay back the principal at the time agreed upon.

These are the costs of production. There are also the costs of selling and delivering what Henry manufactures. Henry must charge more than these costs if he is to make a profit. Otherwise, his capital is making money for others and not for himself.

Many people in North Carolina invest their money and capital wealth somewhere else. And the wealth to start new factories, new industries, new business in North Carolina often has to come from outside the state. The interest on this money is of course paid outside the state, to whoever lends the money.

North Carolina receives many benefits from this capital wealth, such as employment for people, markets for raw materials, taxes paid to the state and local governments. But if more of the profits remained in the state, if people in North Carolina had more money to lend for starting new business, North Carolina would have greater capital wealth.

Capital wealth produces more capital wealth, more of all the different kinds of wealth. All sources of wealth depend upon one another. When capital wealth increases, technological and institutional wealth increase, and natural and human wealth are put to better use.

These Activities Will Help You Understand Capital Wealth Better

1. List ten or more actual examples of capital wealth in your community.

2. Consider what capital wealth you would need in making a home or school garden. In canning fruits or vegetables. In setting up a shop to make wooden toys or chairs. How could you go about getting the capital needed?

3. Have a series of two-minute talks to show how capital wealth comes from natural wealth in the beginning, how it helps develop natural wealth, how it is affected by lack of technological wealth, etc.

4. Think about and discuss the meaning of these statements:
"All capital comes from saving."
"Money buried in the ground is not really capital."

5. Try to find out whether the big factories or largest buildings near you are owned by local citizens.

14. *Institutional Wealth*

THE story of man is the story of his effort to make life better and more comfortable. He wants plenty of good food to eat. He wants plenty of clothes to wear, clothes that are clean and that are warm in winter and cool in summer. He wants a better house to live in.

And this is not enough. Man wants more than the things which give him a strong and healthy body. He wants to know more, to have a better understanding of himself and the world he lives in. He wants religion and music and art and all the other things which give strength to his spirit.

Man begins by wanting these things for himself and his family. Then, as he grows in sympathy and feeling, he wants other people to share in what he has.

During centuries of struggle man learned that it was easier to get what he wanted if he worked with others for the common good of the group. So men joined together in setting up agencies to provide services which everybody needed. These agencies which men made for themselves are our institutional wealth.

Where do we find this institutional wealth? It is in our schools, colleges, and universities; in our churches; in our libraries; in our art galleries; in our symphony orchestras and opera companies; in our hospitals; in the governments of our towns, counties, states, and nation. All these are a part of our institutional wealth.

The measure of this wealth is found in the answers to several questions. How many agencies are there, helping man in his struggle for something better? How many needs do they fill? And how many people do they reach?

In education, for example, it is not enough to have schools for all the children in the state. They must be good schools, and they must be equally good for white and Negro and all other races. The buildings must be well designed and well constructed. There must be carefully chosen books in their libraries and all necessary equipment in their laboratories. The courses of study must be well planned. The teachers must be well trained.

The kind of training and education which teachers give depends on the training and education they themselves have had. What, then, are our colleges and universities like? Are they, too, well equipped with libraries and laboratories and all the tools of learning? Are the members of the faculties well prepared?

Do the courses of study give young men and women a better understanding of the world? Are students ready for the responsibilities waiting for them as citizens in their own communities and in the nation?

These are some of the ways we measure the value of our schools, colleges, and universities.

What of the other kinds of institutional wealth?

Are the churches a part of all communities?

Are there public libraries in North Carolina within reach of everyone looking for a book to read?

Does everyone have an opportunity to see good paintings and hear good music?

Are there enough hospitals over the state? Does every-

The government which we have made for ourselves is an old and valuable part of our institutional wealth.

one receive the medical care he needs, not just what he is able to pay for?

The people have set up certain agencies which are intended to look after the general welfare of the public. Are there enough of these agencies? Do they have well-trained staffs? Is there help for the crippled, the blind, the deaf, the speechless, and the disabled—for all who cannot take care of themselves? Are there free clinics to protect people from disease? Are there standards of cleanliness for all places which might spread disease—such as restaurants, grocery stores, butcher shops, dairies, bakeries, barber shops, hairdressers, drug stores? Are there standards of health for the people working in all such places?

North Carolina's institutional wealth is greater in some ways than in others—greater in schools, for example, than in public libraries or hospitals, or music or art.

Yet the state is not satisfied to have its schools better today than they were fifty years ago. The people want even better schools. They want books at every crossroads and everyone able to read them. They want to hear more music. They want more of all the good things of life.

Individuals often use their capital wealth to create some of the agencies which make up our institutional wealth.

A man who is interested in the control of disease builds a large laboratory where scientists give all their time to study. Or he may establish a hospital for those who need

The North Carolina Symphony Orchestra makes it possible for many people to hear music and to play music.

treatment. Another man sets aside some of his capital wealth for the support of an orchestra.

A woman collects beautiful paintings and builds a gallery where anyone may see them. Her son founds a library of rare books, where scholars from all over the country may study. Her neighbor gives books to the library of a country town.

Someone else gives an office building to a university library. The income from this building is used for buying books which are read by students from all over the state, from all parts of the country.

These are some of the ways private wealth is used to create public wealth. Such institutional wealth becomes a part of man's long struggle for a better and more comfortable world in which to live.

All citizens of North Carolina, whether they are rich or poor, have a share in creating the state's institutional wealth. The representatives they elect to the state legislature decide what taxes are to be paid and what these taxes are to be used for. When people vote, they try to choose representatives who will support the best public interest and will spend the public money for the public good.

Everyone in North Carolina, whether he votes or not, contributes to the funds which are used for institutional wealth. Everyone pays taxes. Sometimes people pay taxes directly to the state. People also pay taxes indirectly, whenever they buy articles from anyone who is taxed.

People help to create institutional wealth in many other ways. They buy Christmas seals, and their money helps in the fight against tuberculosis. They buy Easter seals, and their money helps crippled children. They give to the March of Dimes, and their money helps anyone who has been ill with infantile paralysis. They join the Red Cross and other organizations for public welfare. They join a symphony society which supports an orchestra giving concerts in different parts of the state. They join an association which encourages artists.

Year after year, century after century, men work and

How many people living in the state have an opportunity to visit art galleries and see some of the paintings of old and new artists?

plan to have more of the good things of life. Yet even as they have more, somehow they want still more.

Man's idea of a better and happier life is something like that point on the horizon where the road seems to meet the sky. It is always out of reach, always a little farther on.

"We must have certain things," men say to each other. And they agree in general on what these things should be.

"When these things are ours, we will have what we want," men say.

Yet when they reach the goal they set for themselves, when they possess the things they agreed on, they find that they want other things which are beyond their reach. They have other desires which are unfulfilled.

Every year North Carolina's institutional wealth increases. We learn more. We understand more. We are a little nearer to the kind of life we want. And yet, with every gain, our goal moves a little farther away.

First Find Out

1. What different agencies in your community make up its institutional wealth? You will probably discover some you do not already know about.

2. What special services does each agency perform? To help find out about the new institutions you discovered, you might send a committee to visit them and report to the class about their activities.

Then Think and Talk About

1. How to measure the value or worth of libraries, churches, hospitals. Is there some kind of yardstick for these as there is for schools and colleges?

2. Whether the people in your community are using fully the institutional wealth they already have. For instance, if there is a public library near by, how many of the people in your community have a library card?

3. What improvements are needed in your school to increase its value as institutional wealth?

Then Plan and Do

1. Something to help students in your school take better care of school property, use the library more, take better advantage of their school opportunities.
2. Something to help the people in your community know more about their institutional wealth.

15. *The Wealth of North Carolina*

LIKE a storekeeper looking over his shelves to see what he has in stock, we have made an inventory of North Carolina's resources.

We have found that the state has an abundance of natural wealth: fertile soil; frequent rains; a long growing season; water power; rivers and coastal sounds teeming with fish; trees and plants and minerals; wild animals, birds.

And there is also great human wealth: people to think and work, people to make things and people to use things.

These resources are fundamental. They are the raw materials for the whole pattern of living. Having them in abundance, North Carolina could also have an abundance of technological, capital, and institutional wealth.

But up to the present, North Carolina has failed to make the most of its resources; it has not produced the wealth that it could.

The measure of a state's wealth is not taken just by listing what it has, or by comparing such a list with the wealth of other states, to see which is richer in one thing or another. This is a partial measure, but it is not complete. The true measure of a state's wealth lies in comparing what it has with what it could have.

NATURAL WEALTH

Fertile soil, frequent rains, rivers, trees and plants, wild animals and birds.

When we compare what North Carolina has with what it could have, we find that the state is not rich in technological, capital, or institutional wealth.

However, we know that we have enough of the two fundamental sources of wealth—natural wealth and human wealth—to produce the other three kinds in abundance. So we are not willing to let things go on as they are. We are trying to find out how we can make our natural and human wealth produce more of the other kinds of wealth, how we can establish a better balance between our resources and their uses.

And there is still the old question which for centuries has kept men restless and has stirred them with ambition

HUMAN WEALTH

People to think and work, hands to make and do things.

and discontent. How can we come nearer to a better way of living? How can we approach that distant place on the horizon where the road meets the sky?

At first, this seems impossible. Natural and human wealth are the raw materials for technological, capital, and institutional wealth; and these in turn make natural and human wealth more productive. This sounds like a closed circle. How is it possible to make a beginning toward something better? Things-as-they-are look like an endless chain. Where can we break into it, to start another chain of things-as-they-ought-to-be?

As soon as we ask these questions, as soon as we decide the world must be a better place to live in, then the circle is no longer closed, and the chain is already broken. The power to change things lies within ourselves. Through our own will, our own decision, we can break the circle which we thought was closed.

In an agricultural state like North Carolina, the first step is plain. We must make better use of the natural and human resources which we have wasted by misusing them or by not using them at all.

For years men took the wealth of the land for granted, treated it as if it were limitless. Only a few of them farmed in a way which would save the soil or keep it fertile. There was always more land to be had cheap.

Today, when fertile new land is no longer cheap or plentiful, it is not profitable to continue the old way of farming. Men now conserve the acres which are already cleared.

Abandoned fields must be restored to fertility. Farmers must plant the crops most suitable to their land, and these are not always the crops which have been planted in the past. With new crops there will have to be new methods of farming and marketing. There must be a better balanced agriculture, which will make a man's daily work yield more than the simplest necessities of life.

TECHNOLOGICAL WEALTH

CAPITAL WEALTH

Soil and water must be conserved. Erosion must be controlled. Hillsides too steep for the plow must be returned to the forests, or planted with fruit trees, or covered with grasses for beef cattle and dairy herds.

These are a few of the ways to stop wasting natural wealth. Study and experiment will discover many others.

Who is to make these discoveries? This will be the job of the men and women of tomorrow. So our second step toward a better world must be better training, better education, for the young people of today. Technicians, skilled workers, engineers, scientists, research workers—North Carolina needs them all. But there must be more than "know how." There must be "know why," too. Schools, colleges, and universities must offer their students the disciplines of both mind and muscle.

These, then, are our first steps—better use of what we have; better education and better training for everyone.

After these beginnings the other steps will soon follow. Proper use of resources will mean new industries for a state where there are workers trained to understand and control the complicated processes of manufacture. Better planning in agriculture and industry will make a man's daily work worth more. Eventually there will be more wealth for better schools and colleges, better laboratories, more hospitals, more books and music, better government.

These changes, which at first seem like separate steps, all depend on each other, with everything combining to make one great forward movement. Everyone must take part in such a movement or it will get nowhere. Every man and woman, every boy and girl must have a share in it.

Progress cannot be the work of any one group. It must have the support of the young and the old; of whites and of Negroes; of rich and poor; of men in professions and men on farms; of businessmen, factory workers, and workers everywhere; of women behind counters and women at desks; of women at cook stoves and bridge tables; of women who swing hoes and women who swing golf clubs; of fisher folk and mountain folk; of people living in towns and cities and people living in the country.

This does not mean an easy life for any of us. It will not do away with the necessity for work. The lazy and the shiftless will not be fed at the expense of the wealthy and the industrious. But everyone will be given the right of all free men—the right to work.

The "can nots"—those who are not able to work—will be taken care of. The "will nots"—those who have seemed too lazy or too inefficient to keep jobs—may become different people when they have healthier bodies, better trained hands, and better educated minds. The "do nots"—those

Hospitals *Government* *Schools—Libraries*

INSTITUTIONAL WEALTH

who have not been able to find work—will have a chance to show what is in them.

Our way of life is an old one, which people have followed for many years. Changing it will not be easy. It cannot be done by wishing. There is no magic to bring it about. We must plan. And we must work.

And although we make plans, we must be ready to remake them from time to time. There is no one plan of which we can be sure in a world which changes not only with the years but overnight.

We cannot help making mistakes, because things will not always happen the way we expect them to. What seems to be the solution of a problem may create other problems many times more serious. And what happens in other states and other countries will change conditions for us.

Art galleries *Orchestras* *Churches* *Parks*

North Carolina is part of a larger region with similar resources of natural and human wealth. Virginia, North Carolina, South Carolina, Georgia, Florida, Alabama, Mississippi, Louisiana, Arkansas, Tennessee, and Kentucky—these eleven states of the southeast have many things in common. In the lay of the land, in climate, in soils, in agricultural methods, in special cash crops like cotton and tobacco, in educational opportunities, in racial inheritance, in patterns of living, in loyalty to each other, in countless ways these states are much alike.

Having an abundance of natural and human wealth, the region has not yet made the best possible use of them. The states of the southeast are also much alike in the waste of their resources.

We in North Carolina need first of all to realize what we have and what we are doing with it, to understand what

BETTER TRAINING FOR EVERYONE

we might be able to do, and to know the job ahead of us. We will then be able to work with our neighbors in the region for the common good of all of us.

Generations have passed since North Carolina was a wilderness which brave men set out to conquer, and living is no longer a lonely struggle against an untamed, unfriendly country. Yet there are frontiers in North Carolina today, and we still fight to make this land our own.

Now Is the Time!

To Check Up on What You Have Learned

1. Have a general review of all words and terms learned. You might want to have this review in the form of an

The Arts *The Home* *Science*

old-fashioned spelling match, or you might prefer to be more modern and do it by means of a word puzzle.

2. Have a general oral and written review of the facts learned from this book and from your other activities. The oral review could be carried on like a radio quiz program. Make a balance sheet of the state's wealth to put in your notebook.

To Have an Exhibit

Bring together all the collections you have made and the various booklets, posters, drawings, etc., you have prepared while studying about the state's five kinds of wealth. Arrange and label these in such a way that other people can see their connection and meaning. Later select from these things and exhibit those items of more lasting value and interest. Use them to start a permanent North Carolina museum in your school. Or add them to the museum, if your school already has one.

To Share with Others What You Have Learned

1. Put in writing some of the hopes and plans you have for North Carolina, now that you know more about what your state has and what it needs. You might write an editorial, a story, a skit, a poem, or a musical composition. Take steps to get the best of the writings done by your class published in the school or local paper.

2. Have a program to which you invite your classmates, parents, and others in your community. As a part of the program tell about your work so far, and explain the exhibit you have arranged. You might want to read some of your poems or editorials, and then end the meeting with a party.

To Look Ahead

1. Think about:

a. How you, now and later, can fit into the program of things-as-they-ought-to-be in North Carolina.

b. Whether you are learning the "know why" as well as the "know how" the authors are talking about.

c. How well you are developing the habits and the attitudes frontiersmen of the future will need. Remember—"What you are to be, you are now becoming."

2. Turn through the remaining pages of this book to see what they are about. What are the titles of the other big parts? What are the topics dealt with in each part? Do you see, even before you actually study these chapters, how they carry on the ideas so far discussed?

III. CREATING WEALTH BY FARMING

16. *An Agricultural State*
17. *Farming as a Business*
18. *Small Farms, Many Farmers*
19. *What North Carolina Grows*
20. *The Farmer's Year*
21. *Tenant Farmers*
22. *Sharecroppers*
23. *Farm Owners*
24. *North Carolina Needs More Livestock*
25. *North Carolina Needs More Diversified Farming*
26. *Better Use of the Land*
27. *Working Together Toward Better Farming*

Ever since North Carolina was settled, men have been working close to the soil.

16. *An Agricultural State*

THE first permanent settlers in this "fair and spacious Province" were farmers—families from farther north who came to the land above Albemarle Sound in search of fertile soil. From that time on, forests gradually gave way to clearings, and plowed fields spread westward even into the mountain valleys, until farms took possession of almost two-thirds of the land.

Altogether there are about two and a half million rural people in North Carolina. More than one and a third million of these people—about four in every ten of the state's population—depend entirely on farming for their living. The rest of the rural people are not farmers, but they live in the open country or in villages and towns too small to be counted as cities.

In this country anyone is free to choose the work he will do in order to make a living. Some men farm, just as other men do other things.

When a man decides to become a farmer, he hopes to have more than a place to sleep, enough food to eat, and clothes to wear. He generally hopes for the freedom and independence which belong to any man who owns the land he plows and plants. He knows that a farmer working his own fields is more completely free than the farmer working land which does not belong to him.

North Carolina has been a farming region ever since the first permanent settlers started clearing the land for their crops.

Like anyone else, a farmer wants to have a comfortable house, proper food, and other essentials for himself and his family. In addition to the necessities of life, he hopes for something more—the feeling that he is his own master.

Are the farmer's hopes fulfilled? Is farming a successful business in North Carolina?

This is a question of importance to everyone in the state. Naturally it is most important to farm families, because they are the ones directly concerned in the answer.

But people who have no actual connection with farming are also interested. There are unselfish reasons for this interest. Men working together for the common good want security and the best possible life for everyone. And there are selfish reasons, too. Unless farmers have money to spend, they are not able to buy things which other people must sell in order to make a living.

To find out how successful our farming is, federal agents who take the Census of Population also take a Census of Agriculture. They ask farmers many questions suggested by agricultural experts. In addition, North Carolina has its own agencies which collect information every year about farming in the state. All these facts give us a general idea about farming in North Carolina and the nation.

When we compare North Carolina farming today with that of ten or twenty or fifty years ago, we see that our farming has changed in many ways. When we compare North Carolina farming with farming in other parts of the country, we find that in some ways North Carolina ranks high and in other ways it does not.

Specialists study the agricultural records of the past and the present in order to find out what the future of farming in the state can be. To them, the facts in the records are like signposts at the intersection of several highways, showing where the different roads lead to. The specialists

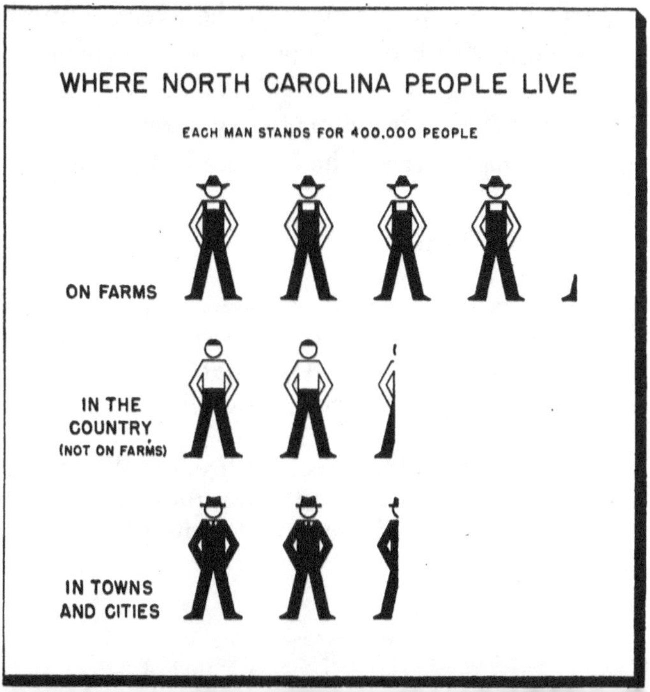

can read these signs and tell the farmers what they mean. But the farmers must decide for themselves which way they will go.

When people discuss agriculture, they sometimes talk about the average farmer, or the average farm, or the average farm income. Just what does this "average" mean?

Suppose that during a year five farmers make an income of $500 each, five other farmers make $1500 each, five more make $4,000, and five more make $5,000. The average income of these twenty farmers is $2,750. Yet there is not a single farmer among the twenty who has exactly this income. It is the "average."

Who is the average North Carolina farmer? He has never plowed a furrow, and he is not to be found in any of the state's hundred counties. It is only on paper, in charts and tables of statistics, that we can find an average farmer, with a family of average size, living in an average farm house, on a farm with the average number of acres, working with the average number of tools and machinery, and earning an average income. But even though the average farmer is an imaginary person, we learn something from him about all the farmers in the state.

Keep In Mind

1. That farming affects *directly* around forty per cent of all the people in North Carolina. Show in a circle graph what part of the total population in the state lives on the land, and what part makes a living by farming.

2. That farming affects indirectly all the other people in the state. Give as many examples as you can of how non-farmers are affected by farming. Read Chapter 1 in

Chapman, *Successful Farming In The South*, and Chapters 24, 31, 32 in Hammonds and Woods, *Today's Agriculture*, for ideas about the importance of farming.

3. That farmers, as well as industrial workers, lawyers, teachers, doctors, bankers, and others, want good living conditions, good opportunities, good fun, etc. Imagine you are a young man on a farm trying to decide whether to stay on the farm to make your living or leave it to work somewhere else. Write a story or skit to show what thoughts would run through the young man's mind as he considered the advantages and disadvantages of farming as a business. After you have finished the study of this entire unit about farming, write another skit or revise this one if your ideas about farming as a promising business have changed.

17. *Farming as a Business*

IN 1945, according to the Census of Agriculture, there were 287,412 farms in the state. Only about half of these—144,450 in all—were owned in full by the men who farmed them. There were 19,835 part owners—men owning some of the land they farmed and renting the rest. And there were a few managers, men who superintended the work on places owned by others. All the other farms—more than four out of every ten farms in the state—were worked by tenants who paid rent in one way or another to men who owned the land.

Some men are tenant farmers from choice. But if they are successful, few of them wish to remain tenants. Like farmers everywhere, most of them hope to own their land.

Then why are four out of every ten farmers in the state tenants? The reason is generally this: they probably do not have enough money to buy farms of their own. By the yardstick of farm ownership alone, farming has not been a success for four out of every ten farmers in the state.

In some statistical tables for agriculture, North Carolina ranks high. We rank high among the states in the total value of our farm products. But we have this high rank mainly because we also rank high in farm population. In the number of people on farms we are second in the nation.

The table on page 125 shows how much cash was taken in by North Carolina farmers for the products they sold in various years back as far as 1925. We see in this table that the cash income in agriculture has been increasing. In 1945 the amount was much more than double the amount in 1925. So North Carolina agriculture is making progress.

But when we read the figures in this or any other similar table we have to remember one thing. We must not forget that in 1945 a dollar did not buy as much as a dollar bought back in 1925.

A dollar always looks the same year after year, and it continues to be equal to one hundred cents. Yet the actual value of a dollar is not always the same because it buys more in some years than in others.

In 1925, for example, a pound of creamery butter cost about fifty-five cents—more than half a dollar. In 1935 and 1940 it cost thirty-six cents. In 1943 it was worth fifty-six cents again. And in 1946 there were months when one pound of butter cost more than a dollar.

So whenever we talk about the amount of money made

by farmers or teachers or workers in factories and offices or anyone else, we must always ask, "How much will this money buy? How much is the dollar worth?"

CASH INCOME FROM FARM PRODUCTS SOLD IN NORTH CAROLINA

1925	$278,685,000
1930	182,694,000
1935	217,259,000
1940	201,241,000
1941	283,102,000
1942	424,206,000
1943	482,544,000
1944	611,415,000
1945	648,777,000

Of course it is not possible to measure the success of any business entirely by the amount of money it takes in. Farmers, for instance, have several advantages which never appear in their bank balances. For one thing, they get much of their living right on the farm. Yet cash income is one way to make an estimate of farming as a business.

How much money, then, do farmers receive each year for the things they sell? In 1941, farmers in North Carolina had an average cash income of $1,037 from all products sold. Because of the war, farm incomes all over the country became much higher than they had been. And in 1944, according to an estimate made by the United States Department of Agriculture, farmers in North Carolina had an average cash income of $2,197 from all products sold.

This amount was much higher than the income in 1941. Yet it was much less than the average in most of the other states. Farmers in thirty-seven other states had a larger average cash income from all goods sold. And the average for farms in the whole country was about $1,050 more than in North Carolina.

In 1945 almost fifty out of every hundred farmers in North Carolina had automobiles. Sometimes the car is better than the house.

A farmer's income can never be all profit. And when this income is very low, he may find that he has taken in less money than he spent. Out of his income the farmer has to buy seed and fertilizer and pay other costs of producing his crop. In 1944, when the average cash income on the farms was a little less than $2,200, the average cash expenses were about $900.

In 1944 North Carolina was thirty-eighth among the states in average cash income per farm for everything sold. We have never ranked high in this. But for a number of years North Carolina has ranked first in the nation in the amount of money spent for fertilizer.

In 1939 the fertilizer bill for the state was over twenty-six and one-half million dollars. In 1941 it was about thirty and one-half million dollars. In 1945 North Carolina farmers used 1,512,000 tons of fertilizer. At thirty-two dollars a ton, it cost forty-eight million dollars.

Every year North Carolina farmers use more and more

fertilizer. In most years the cost of this fertilizer equals about one tenth of the total farm income.

The farmer also has to pay for repairs and replacement of machinery and tools. He has to pay for work animals, taxes, clothing, medical care, and other expenses. And he has to buy anything which he cannot—or does not—produce for home use.

Even though he hasn't a great deal of money, a farmer may have many of the things which must be bought by people who do not live on the land. Much of the food for himself, his family, and the farm animals may be grown on the place instead of being bought at a store. This is one of the advantages which do not appear in the farmer's cash income but which may show in his bank balance.

But many of our farmers do not produce their food and feeds. And this is one reason why farming in North Carolina is not more successful than it is, either as a business or as a way of life.

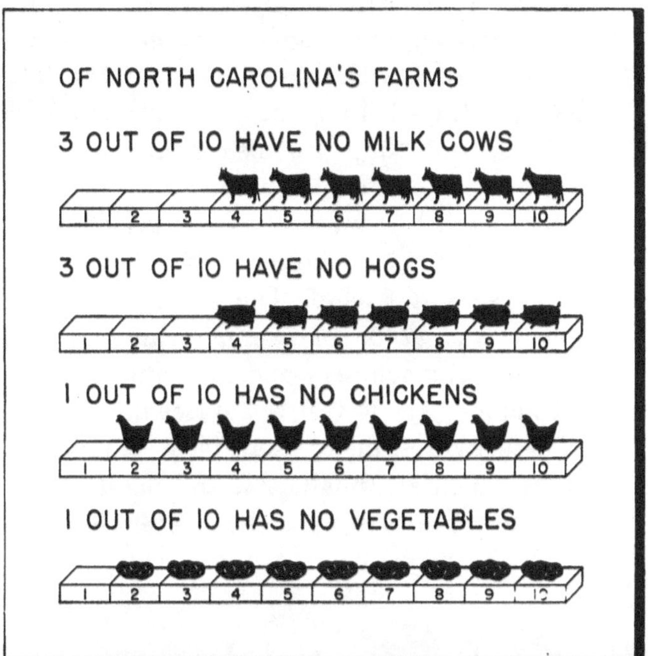

In 1945 over a third of the farms in the state—more than 100,000—had no milk cows. About 95,000 farms had no hogs or pigs. When the farm census was taken, more than 43,000 farms reported no chickens or other kinds of poultry. And more than 24,000 reported no vegetables grown during the year before.

The value of what was produced for use at home averaged only $360 per farm in 1944—not quite a dollar a day. This is not very much for families averaging about five persons. It leaves a great deal to be bought, if the farm family is to have what it needs.

But since the farmer's cash income is small, he is not able to buy a great deal, and he and his family often do without many necessities. They are not getting the full benefit of farm life when they live without fresh meat, milk, and butter, without chicken for Sunday dinner, without green vegetables on the table, without eggs for breakfast and red ham gravy on their biscuits.

What is the value of the farms, farm buildings, and farm machinery in North Carolina? And how do these values compare with those in other states? The answers to these questions help us make an estimate of farming as a business, and they are also a partial measure of our technological and capital wealth.

Here again the state does not rank high. In 1945 the average value of all farms in North Carolina—land and buildings—was $3,440. The average value of farm machinery was $260. In average value of farm land and buildings, of machinery and farm animals, North Carolina ranks among the lowest ten states in the nation.

These figures mean that farm families often live in houses which are small, old, and poorly built. Farm animals do not always have the shelter they need. Hay and corn and other crops are not always protected from rain or sun. Farmers do much of their work by hand, with little equipment. Their few necessary tools, like plows and harrows, are frequently left exposed to the weather in the corners of the fields or under open sheds.

In 1945 almost exactly half of the farmers had automobiles. About 37 per cent of the farms had electricity, about 7 per cent had telephones, and about 15 per cent of all the farm houses in North Carolina had running water.

People can live happily, if not so comfortably, without being able to wash the supper dishes at a sink with running water, telephone a neighbor, switch off the electric lights, and drive over in the car to see friends down the road. Many have managed to do without these conveniences in the past, and some of them will probably continue to get along without them for many years to come. But modern improvements are useful, and people today generally want them—farmers just as much as everyone else.

North Carolina farmers buy these conveniences as soon as they can afford them, and perhaps before. Even so, according to this yardstick, farming in the state does not rank very high.

These statistics about our agriculture would be discouraging except for one other fact, the most important fact of all: farming in North Carolina can be made a more successful business and a more satisfactory way of life.

We have the resources which are necessary for agriculture. Why have we made no better use of them? For years a great many people in North Carolina have been raising crops. Why is our farming no more successful than it is?

What Are the Facts

1. About the number of North Carolina farmers not owning farms?

2. About the value of all farm products in North Carolina?

3. About the average cash income per farm?

4. About the amount spent for fertilizer?

5. About the value of products produced for home use by farmers?

6. About the value of farm buildings and equipment?

7. About the rank of your county according to these yardsticks?

Why not make a bar graph or some type of chart to show the comparisons in each of the facts in 1–7?

What Are the Reasons

1. Why North Carolina ranked eighth in value of *all* farm products in 1944 but thirty-eighth in average income per farm?

2. Why a farmer's income can never be all profit?

3. Why so many North Carolina farmers do not produce their own food and feed?

4. Why North Carolina spends so much on fertilizer?

5. Why the value of farm buildings and of farm equipment is in some ways a measure of technological and capital wealth as well as of success in farming?

If you do not know the answers to all these questions, try to learn them by the end of this part of the book.

18. *Small Farms, Many Farmers*

HERE are good soils, enough moisture, and a long growing season—the essentials of agricultural plenty. Here are people—wealth of man power. Here are consumer markets both at home and within shipping distance—essential to agricultural wealth. These natural and human resources are not being used to the fullest extent possible. They are being wasted.

Why is it that farming in North Carolina does not produce a larger income and a better way of life?

The answer to this question is complex and confused, and cannot be written down briefly. We cannot always be sure what is a cause and what is an effect, or where one leaves off and the other begins.

For one thing, North Carolina's large farm population is crowded into a small area of land. There are only about nineteen million acres of land in farms, with only a little over six million acres in harvested crops. But there are a great many farms—more than 287,000 in all. In order to have such a large number of farms on such a small amount of land, many of the farms are necessarily small.

In 1945 the average size for all North Carolina farms was a little less than sixty-five acres, and more than half of the farms were even smaller than this small average. There were more than three thousand farms in the state with less than three acres, but less than four hundred farms of over a thousand acres.

Not all of the land in farms is being used for crops. This would be impossible in a state which has both mountains and swamps. Some of the land is too wet or too hilly to be cultivated. Some of it is covered with trees. Some is in pastures, and some is lying idle.

Some of the land, abused and worn out, has been abandoned by farmers. Much of this abandoned land now has old field pines as its only crop. Or it is being cut into gullies by erosion, after years of being cut into furrows by the plow.

With more than one and a third million people in North Carolina entirely dependent on farming for their living, only one fifth of all the land in the state is being used to produce that living. The average amount of land being cultivated on a farm is twenty-one acres. And about two-thirds of North Carolina farmers actually use less than this.

In some parts of the world and in some of the other states in this country, a fairly large average cash income is produced on farms which have a small amount of land under cultivation.

In Florida, for example, the average amount of land cultivated in 1944 was about thirty acres per farm, and the average cash income from goods sold was $5,404 per farm. What does Florida grow to produce this income? The mild winters make it possible for farmers there to send fresh fruits and vegetables to market when the weather is still too cold for such crops farther north.

In Connecticut and in Rhode Island, where farms had an average of around nineteen acres under cultivation in 1944, the average cash income from goods sold was $5,445 and $4,627 per farm.

DOES NORTH CAROLINA HAVE TOO MANY SMALL FARMS?

In 1945 there were

3,187 farms with less than 3 acres
105,235 farms with 3 to 29 acres
95,368 farms with 30 to 69 acres
55,846 farms with 70 to 139 acres
10,888 farms with 140 to 179 acres
9,176 farms with 180 to 259 acres
5,813 farms with 260 to 499 acres
1,522 farms with 500 to 999 acres
377 farms with more than 1,000 acres

In Massachusetts, where farms had an average of about seventeen acres under cultivation in 1944, the average cash income from goods sold was $4,403.

In these three New England states farmers sell their products—dairy products, fruits, vegetables, chickens, and eggs—on some of the same large city markets which pay high prices to Florida farmers at other times of the year.

In North Carolina, however, where farms had an average of about twenty-one acres under cultivation in 1944, the average cash income from goods sold was only $2,197 per farm.

In some years this average income is more than it was in 1944. It was higher in both 1945 and 1946, which were the best years North Carolina farmers have ever had. But in many years the farm income is much less than it was in

1944. In 1940, for example, the average cash income from goods sold was $790 per farm in North Carolina.

Farm prices vary with weather and markets. They vary with the conditions in manufacturing and other industries. They vary with national and world affairs. And a farmer never knows whether his income for the next year will be more or less than he had the year before.

But one fact does not change—the fact that the average cash income on farms in North Carolina is not as high as it could be.

What is the reason for this? What are the crops that somehow fail to bring more money and greater security to the average farmer in the state?

You'll Want to Know

1. How to explain the difference between the two figures given about size of farms—
   Average 65 acres for all farms
   Average 21 acres in cultivation

2. What is meant by consumer markets.

3. What different things can cause a farmer's income to vary greatly from year to year.

You'll Enjoy

Trying to show in cartoons some of the causes for changes in a farmer's income from year to year. Some years ago a famous cartoonist drew a picture entitled "A Farmer's Regular Boarders." Around the farmer's dinner table—ready to eat up his profit—were such troublesome boarders as crop failure, floods, insects, droughts. In what other way can you represent the uncertainties faced by a farmer?

19. *What North Carolina Grows*

COTTON used to be "king" of North Carolina agriculture, but the total income from the sale of cotton and cotton seed is now only about a fourth as much as the income from tobacco. Tobacco earns more money for North Carolina farmers than any other crop—sometimes more than half of the cash farm income.

Tobacco and cotton together are our two leading money crops. Their cash value is greater than that of all the other major crops sold in the state.

Summer travelers through Carolina, noticing how often they pass cornfields along the roads, might think that corn is our biggest crop. And so it is, in the amount of land it occupies. But corn as a source of cash income ranks low, because very little of it is sold.

Most North Carolina corn is used on the farms where it is grown. It is made into cornbread for the families. And it feeds the animals on the place—especially the horses and mules which are needed in producing tobacco and cotton, and the hogs which are raised for meat.

Peanuts are important among the minor crops. Pecans and other nuts are grown in small quantities.

There are many different kinds of hay: lespedeza, which in 1940 was the most important hay crop in the state; alfalfa; clover; timothy; grains which are sometimes cut green for hay; and wild meadow grasses. Some soybeans, peanuts, cowpeas, and sorghum are used as hay.

In addition to corn, farmers raise other grains—wheat, oats, rye, barley, and buckwheat.

There are berries and fruits, such as apples, blackberries, blueberries, cantaloupes, dewberries, figs, grapes, peaches, pears, plums, strawberries, and watermelons.

There are various kinds of vegetables, which are often grouped together as "truck crops": artichokes, asparagus, butter beans, beets, cabbage, carrots, cauliflower, celery, chard, collards, cowpeas, cucumbers, eggplant, English peas, Irish potatoes, kale, lettuce, mustard, okra, onions, parsley, parsnips, peppers, pumpkins, radishes, rape, spinach, squash, string beans, sweet potatoes, tomatoes, and turnips.

The variety of these crops is great. Their cash value, however, is small, because they are not produced in large quantities. They are only minor crops.

The table on page 137 shows the cash value of the chief crops in 1944. It is easy to see from this list that our farmers are raising a great many different things. And we can also see how important tobacco and cotton are, compared to the other crops.

The amount of money produced each year by the different crops is sometimes more than the table shows and sometimes less. Up to 1944 it was always less—usually much less. And it was more than this in 1945 and 1946 because everything cost more and was worth more in the years after the Second World War.

While the value of each crop varies from year to year, the share of the total income produced by each crop is generally about the same. Tobacco and cotton are always the main sources of income.

In this table we also see the importance of field crops in the state. Tobacco, cotton, grain, fruits, and vegetables yield most of our farm income. They are all field crops. Livestock is at the bottom of the list.

CASH VALUE OF CHIEF CROPS AND LIVESTOCK SOLD IN NORTH CAROLINA IN 1944

Tobacco	$334,250,000
Cotton and cotton seed	80,590,000
Peanuts	25,449,000
Corn, wheat, and other grains	15,602,000
Potatoes, sweet	6,560,000
Potatoes, Irish	5,274,000
Truck crops	9,352,000
All fruit crops	10,486,000
Cattle and calves	11,293,000
Hogs	24,780,000
Dairy products	26,081,000
Chickens, broilers	20,662,000
Eggs and chickens	20,962,000

For years farmers in North Carolina have been depending on field cash crops for their living—even to buy much of their food. Raising livestock has had a very small part in their farm programs.

In 1920 North Carolina ranked next to last among the states in value of all livestock per farm. According to the Census that year, the average farm in the United States had about three times as much livestock as the average farm in North Carolina. In 1930 North Carolina ranked forty-sixth in value of all livestock per farm.

And what about 1940? Production of milk, butter, meats, chickens, and eggs had increased. Yet in 1940

Corn is raised in every county in North Carolina, but very little of the crop is sold. Most of it is used on the farms where it is grown, as food for the families and feed for the animals.

North Carolina ranked only forty-fifth among the states in value of all livestock per farm, and two-thirds of that value was in work animals. Since then, farmers have raised more livestock every year, both for their own use and for sale.

Farming in North Carolina gives the impression of being very successful when tables of statistics leave out animal production and do not give the average income per farm. In 1940, for example, the state ranked fourth in the total value of field crops alone. But in that same year the state ranked thirtieth in the total value of animal products, and it ranked last of all in the percentage of farm income from livestock and livestock products.

When the values of field crops and of animal products were both included—as they should be—North Carolina ranked fourteenth in total value. And the rank in value per farm of all farm products in 1940 was thirty-eighth —much lower than the rank in total value.

North Carolina agriculture has been more one-sided than it needs to be or should be. We have raised crops in large quantities, but we have too often specialized in plants grown in the fields. We have never raised the animals that we could—either dairy cattle or beef cattle or poultry or other livestock. Plants and animals are both necessary for a balanced agriculture in North Carolina.

Animal production is increasing, and plant production is becoming more varied. But many farmers in the state will continue growing tobacco or cotton as their main cash crops for a long time to come. These are the crops best suited to the soil and climate in several sections of North Carolina. And they will probably always have a place in a well-balanced agriculture.

For years many farmers depended entirely on these two cash crops for their living, and tried to make this living better by growing more tobacco or cotton. In doing this, they sometimes used land which was unsuitable, and planted more acres than they could take care of properly. As a result they often had crops of poor quality which brought low prices on the market.

Even when the land was fertile and the crops good, farmers who raised as much cotton and tobacco as possible did not always make a large profit from their crops. When they had crops larger than the markets needed, farmers often received less money than they would have if their crops had been smaller.

In the 1930's Congress passed Agricultural Adjustment Acts which were supported by farmers. Under these Acts

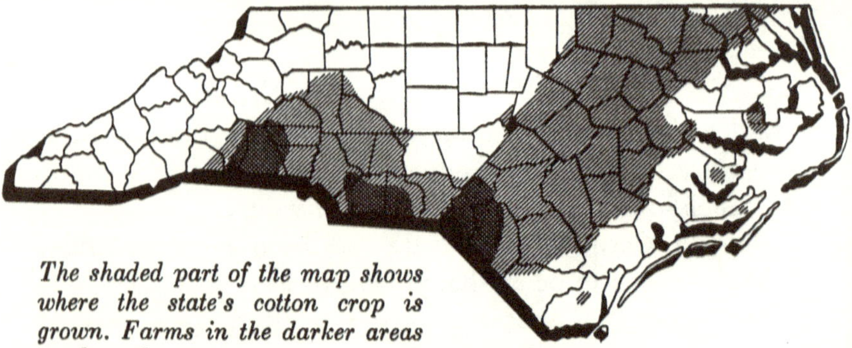

The shaded part of the map shows where the state's cotton crop is grown. Farms in the darker areas produce the most cotton.

THE COTTON BELT IN NORTH CAROLINA

many farmers used smaller amounts of land for tobacco and cotton and planted more soil improvement crops. The federal government helped to pay the cost of improving the soil.

Most tobacco and cotton growers will be better off if they follow some of the suggestions intended to make these two important crops part of a well-planned, balanced farming program.

Sometimes farmers still grow tobacco and cotton on land which is better suited to something else. This misused land should be planted with different crops—other field crops, soil improvement crops, cover crops for pastures where livestock can graze, or trees for reforestation.

Farmers should try for quality in their cash crops. In planting tobacco and cotton, they need to select varieties recommended for their particular soil and climate, and to buy clean quality seed which has been tested for germination. In tobacco, they should try to produce only the best leaf. In cotton, they should try to grow a longer staple. Only recommended fertilizers should be used. And the control of plant diseases and of insect pests should be a part of every farm schedule.

With a program like this, tobacco and cotton farmers

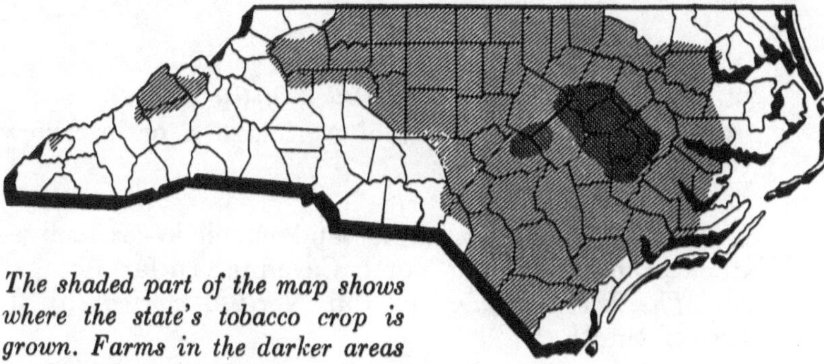

The shaded part of the map shows where the state's tobacco crop is grown. Farms in the darker areas produce the most tobacco.

THE TOBACCO BELT IN NORTH CAROLINA

will have a chance to receive a better income for a year's work. Crops of high quality bring the highest prices.

This income will mean more to farm families if more food and feeds are produced for home use. When tobacco and cotton farmers become live-at-home farmers, all the money from cash crops does not have to be spent for food. It can be used to buy other things for the family, the farm, and the home. And some of it can be saved.

Can You Explain

1. These terms: major crop, minor crop, field crop, cover crop, soil improvement crop, balanced farming, two-armed agriculture?

2. The reasons why North Carolina produces so little livestock?

3. This statement: "Farmers should try for quality in their cash crops"?

Can You Find This Information

1. Chief sources of income from crops and livestock in your own county? See *Agricultural Statistics*, published

by the State Department of Agriculture, or interview your farm agent.

2. Which states rank higher and which lower in livestock production than North Carolina? In farm crops? See *The World Almanac*, and *North America*, by J. Russell Smith.

3. Important facts about farming in other states? See chapters about this in Chapman, *Successful Farming in the South*, Chapters 2 and 3.

Can You Show

1. On a graph the sources of income from farm crops and livestock as given on p. 137?

2. On a graph the same information about your own county?

3. On a map the crops grown in your county?

4. In a table North Carolina's rank among the states according to several agricultural yardsticks?

20. *The Farmer's Year*

AS LONG as tobacco and cotton remain the chief sources of farm income in the state, these two crops will shape the lives of thousands of farmers.

The details of farm life follow a general pattern determined by these two crops—the way a man lives and works, the size of his farm and whether or not he owns it, how large his family is, when the children go to school, what they eat or wear, how healthy they all are.

Take Jeff Bowers, for example. Jeff grows flue-cured tobacco. He has his own way of living and of raising a crop, and it isn't exactly the way every other tobacco farmer in the state lives and works. But Jeff's methods are typical of many farmers in North Carolina who raise a little flue-cured tobacco and not much of anything else.

Tobacco seeds are so small that they would be lost if they were sowed directly in the open fields, and the young plants are so tender that they need protection from cold weather. So Jeff sows the seeds in a plant bed, and later he transplants the seedlings into the open field.

On a clear day early in the year Jeff begins work on the bed for his plants. He is making a new bed this year, and has already chosen the place for it. It is a sheltered slope slanting toward the south for warmth and sunlight, with a

Tobacco cloth is stretched over the plant bed to keep out wind and cold and insect pests.

stream not far away so that water can be carried to the seedlings if there is a long drought.

Jeff's ax bites sharply into the trees and sends them thumping to earth that is still hard with the night's frost. He clears the trunks and drags them aside. He will need them later. The branches he heaps on the ground, adding old logs, leaves, and brush until the pile is perhaps as high as his knees. Then, on a calm day when there is no more than a breath of wind stirring, Jeff touches a match to one corner. A fire that may burn into the night is soon filling the air with drifts of wood smoke. Insects and weed seeds near the surface of the ground are killed by the heat, and the ashes improve the soil.

Jeff digs up the tree roots, removes all rocks and trash, and works over the soil carefully with plow and harrow until it is fine and smooth. Fertilizer is added and worked in.

The tree trunks he had dragged to one side are now placed firmly around the edges like a low wall. The tobacco bed is ready.

Late in January or early in February Jeff mixes the tiny seeds with sand or ashes or cornmeal to keep the plants from coming up too close together. Then he scatters the seeds over the soil and presses them into the earth with a board. To protect the bed from wind and cold and from insect pests like the flea beetle, he fastens a cloth to the bordering tree trunks and stretches it tight a few inches above the bed.

While Jeff is waiting for the seeds to show their first leaves, he cuts wood for curing the tobacco later on. If he will need more tobacco sticks at curing time this year, he cuts them now, splitting them out of straight-grained pine. He mends the roof of the tobacco barn, looks over the walls and fills in any chinks where the mud has fallen out from between the logs, and makes sure that the flues are in order and ready to be used.

Before these jobs are finished, the first leaves of the tobacco plants are beginning to push above ground. On warm clear days Jeff folds back the cloth so that the plants will be exposed to the sun. He watches for signs of insects, and of diseases like blue mold or black root rot. And before the cool air after sundown can touch the plants, he carefully puts the cloth back in place.

Weeds have to be pulled out of the plant bed. And if there are several weeks of dry weather, the seedlings must be watered. Jeff's wife, Sarah, helps with this work.

When the days are warm enough—perhaps in the first part of April—Jeff follows his mule, Major, back and forth across the fields. He breaks the land with a plow, drags it to make the soil fine, and ridges it into long fertilized rows.

As soon as there is no danger of a killing frost, Jeff is ready to transplant. He has taken the cover off the bed so that the plants will not be too tender for the open field. Now all he needs is rain.

Worming and suckering can often be done at the same time.

If there is a long drought, he cannot wait for rain. Then every plant must have a little water poured into the earth at its roots when it is set into the ground. The two older boys may have to stay home from school to help, although their mother doesn't like for them to miss their classes.

Early in the morning the work begins. Jeff chooses strong healthy plants with the dew still on their leaves and the damp earth of the plant bed clinging to their roots. Little Jeff and James Henry walk along the furrows, dropping a plant on top of the hill about every two feet. Jeff and Sarah follow behind. They punch a hole in the soil with a strong lightwood peg, drop in a plant, and press the earth tight around the roots.

Jeff and Sarah work quickly, with the easy skill which comes from many years of practice. Even so, it is a big job, and a backbreaking one. Between five and six thousand plants must be set to every acre.

Four or five days after the transplanting is finished, Jeff

walks through the fields and replaces any seedlings which have been killed by the moving or the cutworms.

A week later he takes Major down the rows with a cultivator, and Sarah and the children chop any grass and weeds which are growing close to the plants. Five or six times during the next two months Jeff works the crop with a plow and the family hoes out the grass and weeds, until the plants are about two feet high and the leaves are so long that they almost meet between the rows.

The plants are beginning to send up tall stalks on which blossoms would flower if they were allowed to grow. Every stalk must be broken off so that the strength of the plant will go into the leaves and not into making flowers and seeds. Breaking out this blossom stalk is called "topping."

Knowing when to top or not to top at all takes judgment that comes only with experience. Jeff learned by working at it with his father, just as his father learned from his own father before him. But there are some things of which the best tobacco farmer can never be sure—the kind of weather ahead, which might make one topping height better than another; and the kind of market ahead, when one sort of leaf may sell better than another.

As soon as the plants begin to mature, they send up shoots between each leaf and the main stem. These "suckers," as the shoots are called, will keep on growing if they are allowed to remain. So they must be broken off.

The plants continue to sprout new shoots as quickly as the old ones are removed, and the job of "suckering" may have to be done again and again. This is work for Sarah, Little Jeff, and James Henry, with the other children helping too. It is work that needs many hands, but the hands do not have to be very skillful or fully grown.

While Sarah and the children sucker the plants, they hunt for hornworms. These dark green worms eat the leaves, and find a diet of tobacco so wholesome that they grow to be four or five inches long. Many farmers get rid of the worms by spraying with poison, but Sarah and the children still pick them off the plants by hand.

Worming and suckering are hot and tiring jobs. The plants grow thick, and the rows seem to stretch endlessly under blinding summer suns. Sarah and the children wear their oldest clothes. And they go home at night with hands and clothes sticky from the sap of the tobacco.

There are other threats to the crop besides tobacco worms. Budworms may eat out the tops of the young plants and keep them from growing. Tobacco mosaic may appear, spotting and twisting the leaves. If there is root knot, it will make the plants small and pale. Granville wilt may attack the crop.

Wet weather generally brings on blackfire or wildfire, which spots the leaves. Warm weather may spread black shank through the fields. Heavy winds may flatten every plant. Hailstorms may cut the leaves into ribbons.

If all goes well, about the middle of July the bottom leaves, or "lugs," begin to ripen. They are "prime"—that is, ready for harvesting. Jeff, like most of the tobacco farmers in North Carolina, harvests and cures his leaves as they ripen.

The first priming begins. Sarah's sister, May, and May's husband, Clyde Robinson, come over with their children from the next farm to help. Sarah and Jeff and their family will help the Robinsons next day, when their tobacco is ready to be primed.

Jeff and Little Jeff go down the rows, breaking off the

The priming begins when the bottom leaves, or lugs, are ready for harvesting.

ripe lower leaves and piling them carefully in the light framework of a wooden "sled." As soon as the sled is full, Major and James Henry take it to the curing barn where the rest of the family is working in the shade of a tree, and Jeff and Little Jeff start piling leaves on an empty sled.

Sarah and May, the "loopers," stand beside racks which hold tobacco sticks about four and a half feet long. Sarah's oldest girl, Catherine, and May's oldest girl, Bertha, are working as "handers." They smooth out four or five leaves at a time and hand them to the women, who quickly loop the stems with string and tie the bunches to the stick with the leaves hanging first on one side and then on the other.

As soon as a stick is full, Lona Bowers carries it to the barn, where Clyde lays it across the poles. Everyone works fast. By the time the sled is empty, Major and James Henry are there with another one piled high with leaves, and by night the barn is closely packed with rows of tobacco ready for curing.

When tobacco is flue-cured, the moisture in the leaves is removed by very strong heat circulating through flues, or pipes. The equipment for doing this is simple.

The barn where Jeff does his curing is about eighteen feet square, built of logs chinked with mud. It has top windows for ventilators, and one door. The floor is of dirt. Poles for the tobacco sticks are fastened from wall to wall in layers about eighteen inches apart, beginning six feet above the ground and reaching all the way up to the roof.

In one wall are two furnaces of bricks, like big ovens, with the openings on the outside under a shed. Inside the barn a large sheet-iron pipe about a foot in diameter is attached to the closed ends of the furnaces. This pipe, or flue, carries away the smoke which otherwise would damage the quality of the leaves.

This is no ordinary smoke pipe. It stretches like a giant worm across the dirt floor of the barn toward the opposite wall, loops around, and then returns to the wall containing the furnaces. There it rises off the ground, passes outside through a hole just above the furnaces, turns up toward the sky, and serves as a smoke stack against the side of the barn.

The flue carries heat as well as smoke, and the lengths of pipe lying across the floor distribute this heat evenly throughout the barn.

Outside the barn stand the piles of pine, gum, and oak

As soon as the leaves have been tied on sticks, they are packed in the tobacco barn. If the barn is already full, the sticks have to be hung on poles to wait for the next curing.

which Jeff cut and hauled in during December and January. This wood is used for curing.

A thermometer hanging just inside the barn door helps Jeff watch the heat of his fires. But there are no set rules as to how hot the barn should be or how long he should keep up the heat. These requirements vary with the leaf itself and with the weather at curing time. If the fire is too quick or too hot, or if the extreme heat is held too long, the leaf is damaged, perhaps ruined. If the fire is not hot enough, if the heat rises too slowly or drops too soon, the leaf is damaged, perhaps ruined.

Only experience can teach a man how to cure tobacco. It is a highly skilled process.

While Jeff is curing, he stays near the barn most of the time. Sarah may bring his meals to him. He naps during the day. And at night he rolls up in a thin blanket under the shed and sleeps for an hour or two. But Jeff does not sleep long. Temperature and tobacco must both be watched.

At last, after about five days, the curing is finished. The fire is allowed to die down, and the door of the barn is opened. The leaves must not be touched while they are dry and brittle; but as soon as the moisture in the air has made them soft and pliable, they are taken out of the barn and stored in the pack house. The barn is ready to be filled again with green tobacco.

These are busy days for everyone. Priming and curing continue as the leaves ripen.

After the lower leaves, or "lugs," are cured, Jeff and Little Jeff prime the center leaves and then the "tips." The different primings are cured separately and are stored separately in the pack house. This makes it easier to sort the tobacco into different grades.

Sticks of tobacco are handed into the barn and hung on crosspieces overhead. A large metal flue pipe winds from the furnace back and forth on the ground. It will carry the heat which cures the tobacco.

By the time Jeff is ready to grade his tobacco for market, the leaves may be dry and brittle again. Then they must be hung in a pit, or in a room where the steam from a steam box will soften them. When the leaves are once more soft and pliable, Jeff says that they are "in order."

In the grading room, which is next to the ordering room, Jeff prepares his crop for market. Carefully examining each leaf, he sorts the tobacco into different piles according to color and texture. Then Sarah and Little Jeff take a handful of leaves—twenty or thirty at a time—and wrap the stem ends with a leaf that ties the tobacco into a bundle. This bundle is called a "hand."

Catherine and Lona hang the hands on tobacco sticks, and James Henry packs the sticks carefully in the trailer. When the trailer is full, Jeff and the boys stretch a couple of old quilts over the tobacco to protect it. It is ready for market.

Everyone goes to bed early that night and gets up early next morning. The whole family is preparing for the special day. Jeff has fed Major, Sarah is dressed and is cooking breakfast, and Little Jeff and James Henry have already carried in some firewood and are sitting, scrubbed and sleepy, waiting to eat. This year, for the first time, Jeff is taking the boys with him.

The stars are still bright when Jeff turns the car from the dirt road into the highway. The boys watch daylight spread across the sky as the early morning smoke of breakfast fires rises from farmhouse chimneys. Later they see men, women, and children going off to work in the fields.

By the time they reach town there are trucks, trailers, cars, even a few wagons—all loaded with tobacco—all waiting to be driven into the warehouse.

Jeff takes his place in line, and at last it is his turn to unload. He slides the hands off the sticks and arranges them carefully according to grade in large flat baskets. Each basket is weighed and is marked with a ticket giving the weight, Jeff's name, and a number.

Then Jeff's baskets are placed in a long row that stretches across a floor as large as Jeff's field of corn beyond the creek. Little Jeff and James Henry help their father cover the tobacco once more with quilts and sacks. When Jeff has parked the car outside, there is nothing more to do but wait and watch.

Later in the morning the tobacco is uncovered and the sales begin. It is a familiar scene to Jeff, but to the boys it is like another world.

Warehousemen and buyers move along the rows of tobacco from basket to basket, stopping for a few seconds while a pile is sold. Not far in the background are the farmers, anxious to know how things are going. And behind them are men who follow the sales because they grew tobacco once and have never lost their interest in it.

A helper pulls several hands from the bottom of a pile to show that there is no poor tobacco hidden underneath.

One of the warehouse operators names a price at which he thinks the bidding should start. And then the auctioneer begins a singsong chant which sounds like a formless tune in an unknown language. He is repeating the price named by the starter.

The buyers do not speak, but by familiar signs they make their bids. A lifted eyebrow, and the auctioneer changes his words and tune to show that the price has gone up. A wink, or a shrug of the shoulder, and the auctioneer's chant changes again. This continues until the tobacco is sold.

Of course the buyers understand what the auctioneer is saying. And so do the men who work around the warehouse from day to day, and Jeff and the other farmers. But to Little Jeff and James Henry it is all a mystery.

As soon as auctioneer and buyers have finished with Jeff's baskets, a helper writes the prices on the tickets, and Jeff goes over to see how much his crop was sold for. Within an hour he is at the cashier's window for his money.

If a man thinks his crop should have sold for a higher price, he can tear up his tickets and put his tobacco up for sale again. He can even load it into his truck and drive to another warehouse or another town. But he has no way of being sure that he will get more. It might be less. Jeff Bowers, like most of the farmers, takes what he gets.

By the time Jeff has graded and sold all his tobacco, his income for a year's work is in his pocket. At least, Jeff thinks of it as a year's work, because it is all he has done during the past year. Yet for weeks in the winter, when he couldn't work on tobacco, he had nothing else to do.

In the summer there were days which were too short for all that needed to be done, but there were days in the spring when Jeff could have worked in a garden if he'd had one. And every day he could have spared enough time to look after a cow and some chickens, if he'd had them. He might better have done this, even if it meant raising less tobacco.

As long as Jeff continues to grow all the tobacco he can, and nothing else, there will be times when he has more to do than one man can manage. And there will be other times when he is idle.

Jeff generally depends on his cash crop for almost everything. And if he loses his tobacco, he loses everything. He takes a big risk.

When he is sick, Sarah and the children look after things the best they can. If it's winter time, it doesn't make much difference. But if it's summer and he ought to be in the fields, he'll lose money on his crop.

Jeff doesn't have any accident insurance for himself, or hail or other crop insurance for his tobacco, the way some farmers do. He'd have to pay for it himself, and he doesn't have enough money for such things.

Of course not every tobacco grower in North Carolina lives and works the way Jeff does. Other farmers sometimes use different methods of doing the same things.

Chopping cotton is a job that takes many hoes and hands.

Young children help with the work when the cotton is ready to be picked.

Some tobacco farmers use hand planters. Some use machines on which men ride as they do the planting. Some farmers no longer top their plants. They get a lighter leaf if the tobacco is left untopped.

A tobacco barn may be built of brick or hollow tile or thick sawed timber, instead of logs. Instead of two furnaces, like Jeff's, the barn may have only one. As fuel for curing many farmers are now using oil or coal. A few are using electricity.

In the western part of North Carolina farmers raise burley tobacco instead of flue-cured tobacco. Their ways of harvesting and curing are very different from Jeff's.

There are many details in which methods of raising tobacco may vary. But Jeff's way is the general pattern of work for the tobacco farmer, especially in the heart of the flue-cured tobacco belt. It is the way thousands of men in North Carolina live and work.

The average cotton farmer follows much the same general pattern of work as the tobacco farmer, although raising cotton is generally thought of as being easier than raising tobacco.

How do the two crops compare in work? Cotton is planted directly in the field. There is no backbreaking job of transplanting, as there is with tobacco. Cotton rows have to be "chopped," like tobacco. But the cotton farmer does nothing so tiresome as worming and suckering, unless it is picking.

When the cotton bolls are open, the cotton does not have to be picked at any particular time, like tobacco. It should not be left in the field too long, but it may be left for a while without serious injury to the crop. When cotton is picked, it is ready to be ginned and sold. There is nothing like the tying, curing, ordering, grading, tying, and marketing which still lie ahead of the tobacco farmer after he has done his priming. These are some of the ways in which cotton is easier to raise than tobacco.

In several other ways cotton growing and tobacco growing are much alike.

1) They are both non-food, cash crops.

2) They both take a large amount of hand labor.

3) Much of this hand labor is done by women and children.

4) Both tobacco and cotton take a great deal out of the soil without returning anything to it. A large part of the

Farmers wait around for their turn at the cotton gin. At the end of the building the finished bales are loaded into wagons and driven away.

commercial fertilizer bought in North Carolina every year is used on tobacco and cotton. But this does not build up the soil, which generally becomes poorer in plant food and humus with every crop it produces. Tobacco and cotton are both clean-culture crops. This means that the soil between the rows must be kept clean of grass and weeds. When this is done, the soil often leaches and washes away.

5) A great deal of the tobacco and cotton in the state is grown by men who do not own their farms. They are tenants.

Do You Agree

1. With the statements in paragraphs one and two of this chapter about the effects on a farmer's life of growing cotton and tobacco? Why or why not?

2. With the general description of tobacco farming given in this chapter?

3. With the idea that cotton farming is easier than tobacco farming?

Do You Know

1. What is meant by clean-culture crops?

2. What effect clean-culture crops have on soil?

3. How tobacco and cotton farming are alike?

4. Why it is that so much of the tobacco produced is raised by people not owning their farms?

21. *Tenant Farmers*

FARM tenancy is a very old method by which men are able to use land they do not own. In the whole United States there is not one state without tenancy.

Farm tenancy is especially common in the South. About half of all the tenant farmers in the nation are in the eleven southeastern states.

And North Carolina ranks high in tenancy. In 1940, only three of these southeastern states had more tenant farmers than North Carolina. Almost half the farmers in North Carolina that year were tenants. And about half of the tenants were sharecroppers. In 1945, about 42 per cent of North Carolina farmers were tenants. Many others were part tenants.

North Carolina tenants may be classified by how they pay their rent and by what they receive in return. Every kind of tenant generally receives a house, land, and the right to cut the wood he uses, unless there are no trees on the place. His other arrangements with the land owner vary.

1) A *cash tenant* pays his rent with money. He generally receives house, land, and wood. Except for his rent, he is entirely independent of the owner. He decides for himself what crops or livestock he will raise.

CASH TENANT *SHARE-CASH TENANT*

2) A *share-cash tenant* pays part of his rent with money and part with the products of his farm—either crops or livestock or both. The crop with which he pays rent must be one which his landlord wishes him to grow. In North Carolina this is generally tobacco or cotton, plus some of the corn crop. The share-cash tenant also receives house, land, and wood.

3) A *share-tenant* pays all of his rent with a share of whatever he produces. He supplies his own work animals, and generally supplies his own machinery and tools, the seed he plants, and part of the fertilizer he uses. The owner furnishes everything else. The tenant and the owner agree on what crops are to be used for rent. The owner generally receives about one-fourth or one-third of the rental crops.

SHARE TENANT *SHARECROPPER*

4) A *sharecropper*—or cropper, as he is usually called—also pays his rent with a share of what he produces. But he pays a larger share than the share tenant because he receives more from the owner. He generally pays half of the crops to the landlord, and sometimes more. As a rule, the sharecropper furnishes nothing but labor and perhaps half of the fertilizer. Everything else is supplied by the owner—the cropper's house, the mules or horses with which he works the crops, seeds for planting, the rest of the fertilizer, the tools and other equipment. Even the money to buy clothes and food for his family is often borrowed by the sharecropper, either from his landlord or from a storekeeper. The landlord decides what the sharecropper plants.

A cash tenant may be free and independent although he does not own his land. Sometimes, even when he has money enough to buy a farm, he remains a tenant from choice. He decides to invest his capital wealth in another way—perhaps a herd of dairy cattle, or other livestock, or some other farm crop.

A sharecropper is less independent and in general has a smaller income than other kinds of tenants. But he has a chance to make a living from the soil without owning any land and without having any more resources than his hands. And sharecropping makes it possible for land to produce wealth even when the owner has no capital with which to pay wages or all the other costs of farming.

Sharecropping may help a young man who is just beginning his farming career. He may receive training and experience while he is saving money to buy a mule, then a plow, then other tools. As time passes, he may become a share tenant, then a share-cash tenant, then a cash tenant, then an owner. Sharecropping has given him a start. This is the way farm tenancy is supposed to be.

When a man uses sharecropping or some other form of tenancy as a step in the ladder to independent, successful farming, then tenancy may have definite advantages.

But many farmers in North Carolina are tenants from necessity because there is nothing else for them to do. Many remain tenants all their lives for the same reason.

A sharecropper is often unable to make enough money to start out for himself, to become a share tenant or a cash tenant, even to own a mule and a plow. He does not climb above the lowest step on the ladder.

Farm tenancy, especially sharecropping, has been criticized a great deal in recent years. Tenancy often takes the place of land ownership and has become a fixed way of life from which many people cannot free themselves.

The Census of Agriculture shows that North Carolina had more farm owners and fewer farm tenants in 1940 than in 1930. In 1945 there were more owners than in 1940, and the number of tenants was a little smaller.

The fact that there were fewer farm tenants in 1940 and 1945 than in 1930 did not mean that the tenants were becoming owners. It meant that many owners stopped renting to tenants, and began farming the land themselves with men who worked by the day.

In 1937, Congress passed the Bankhead-Jones Farm Tenant Act, under which tenants could borrow money from the government for buying farms. This gave some tenants a real opportunity to own their own land. But it was only a small beginning.

Farm tenancy as it exists today must be changed, and the job of changing it must not be left entirely to the federal government. Part of this work must be done by the people of North Carolina and of the South. They must study the facts, find out what needs to be done, and decide how to do it.

So let us look at sharecropping a little more closely. Let us see what it means to the croppers and their families. And perhaps we will also understand what the sharecropping system means to everyone in North Carolina.

Why Not List in Your Notebook

1. The different kinds of tenants and the arrangements each kind has with his landlord?
2. The possible advantages of tenancy?
3. The disadvantages of tenancy?
4. Ways to reduce tenancy?

Why Not Have

Open-Forum Discussions of Some of These Questions?

22. *Sharecroppers*

TENANCY—especially sharecropping—thrives on the same soil and in the same parts of North Carolina where tobacco and cotton grow best. Nine-tenths of all farm tenants in the state are concentrated in the counties growing these two crops. Some of the things which are generally thought of as being disadvantages of sharecropping may instead be disadvantages of growing only cash crops and nothing else.

1) In the first place, the average sharecropper makes very little money. Before prices went up after the Second World War, the cotton tenant in the South had an average income of less than sixty-five cents a day for himself and his family.

During the Second World War and the years that followed, some groups of workers in North Carolina made more money than they had ever made before. Tobacco tenants were among these. They probably had a larger income than any other similar group in the South. But even so, their income was not large.

Tobacco and cotton produce more wealth than any other crops in North Carolina, but very little of this wealth goes through the hands of the average sharecropper. He generally has only a small amount of land planted with his crop, because tobacco and cotton both need a great deal of hand labor. A cropper and his wife and their children are able to take care of only a few acres.

Even if a sharecropper received the money from all of his crop, his income would not be large. But the owner of the land is generally paid half of the crop as rent. Thus the amount of money which the cropper receives for his year's work is often small.

2) The average sharecropper receives his entire income when his crop is sold. He has no money coming in every day, or every week, or every month, as most people do. He receives all his money once a year. In theory, he will use this cash to pay his expenses for the next twelve months, until the new crop is ready.

This is a difficult way for anyone to live. It is hard for anyone to plan a year's spending so well that the family will have what is needed. And it is hard to make certain that expenses will not add up to more than the amount of money a family has. Most people generally find it hard to budget their expenses properly for a week, or a month. To do this for a year is even more difficult.

In average times a sharecropper seldom has a chance to use the money from last year's work to pay this year's expenses. Most of the sharecropper's income is already spent before he receives it, because he has been living on credit during the past year. When his crop is sold, his debts must be paid.

By the time a cropper has paid his debts, he has little if anything left for the next year. So once more he lives on credit, borrowing ahead and promising to pay back the loan with the income from the next crop.

If the crop fails or if it sells at very low prices, the sharecropper does not receive enough to pay back what he owes. Then he must start out already in debt for the year that is past and borrowing for the year that is ahead.

3) The sharecropper often grows very little food for himself and his family. His crop—either tobacco or cotton—may be planted right up to his house. Sometimes he has no time and no land for a garden.

The cropper's wife and children may manage to plant a few rows of beans and peas, and perhaps a watermelon patch. They generally have collards, cabbages, or turnip greens. Sometimes they raise a hog or two for meat. Perhaps they keep a few chickens. And that's about all.

The cropper loses one of the main advantages of farm life when he doesn't grow what he eats. His income is generally too small to buy the right kind of food. There are few green vegetables or fruits on his dinner table. There is often no milk, and no meat other than pork. Too often there are no eggs. Many sharecropper families have little more than meat (meaning fatback), meal, and molasses.

Much of the food and other necessary supplies are furnished the sharecropper as a loan, on the basis of what the land owner or the storekeeper thinks the cropper's income will be. By the time the cropper has paid back the amount loaned for these "furnishings," plus the interest—which is sometimes charged at a high rate—he hasn't much money left.

4) Tenants move often, on an average of more than once every three years. Every year—usually in December, January, or February, before it is time to start spring plowing—North Carolina roads are dotted with trucks and wagons. Over a third of the state's tenant families, with their few pieces of furniture and everything else they own, are on their way to new farms.

The sharecropper's cotton is often planted close to his house. He may have little land and little time for a garden.

Why do they move so often? There are many different reasons. They may be looking for a better landlord. They may move to be nearer a church. They may be hoping to find better land, grow a better crop, make a better living. Some of them may move from habit, because they have already lived on one farm for two or three years, and they are ready to go on.

Wandering from place to place, a sharecropper family has little feeling of security, of belonging to any one community. Many of them miss knowing the neighborly friendliness which people feel when they go to the same schools and churches and talk together about politics and crops, or cooking and sewing. If men and women expect to be moving soon, they take on few of the responsibilities of citizenship, and they are not very much interested in what is happening in the neighborhood around them.

Sharecropper children generally miss several weeks of school work when the family moves. Sometimes, if the family is planning to leave soon, the children do not go to school when it opens. After moving to a new neighborhood, they may stay out of school the rest of the year.

Whatever the children do, the change is an interruption. Some sharecroppers and their families have not gone to school enough even to learn how to read. They are illiterate.

5) Illiteracy is common among sharecroppers.

Tenant children are part of the family wealth—the hands which are one of the great human resources of North Carolina. They help with the crops.

In a section where there are a great many sharecroppers, the school may close when the children are needed in the fields. The school will stay open at other times in the year, so that the school term is just as long as in all the other state schools.

In some counties where cotton is the chief crop, a few schools begin their fall term in July. They close for a while in October, while the cotton is being picked. Then they open again.

In other counties a few schools close early in the spring, so the older children can help pick the strawberry crop. To make up for the time that they have been closed, these schools open again in August.

Even when school is in session, tenant children often stay away. Perhaps they are helping with work at home, or are ill, or haven't proper shoes or clothing. Sometimes, like any children, they just don't want to go to school.

Many tenants—especially sharecroppers—have never learned to write their own names, and many are unable to do much more than this. They are illiterate. Not all tenants are illiterate. But as a group they rank lowest in literacy. This is true especially of the older croppers, who grew up when it was not always easy for rural children to get much schooling. With better schools and stricter laws about attendance, this condition is improving.

What difference does it make to a farmer whether or not he can read and write? A farmer who is illiterate cannot look over the account of what he has bought at the store, or keep a simple record of how much money he spends and how much he takes in. He can't read the instructions for spraying his tobacco plants against blue mold. When he buys fertilizer, he can't read the tag to make sure that it is the kind he wants.

Farm magazines or farm bulletins, with suggestions which might help him plan his work, are of no use to him. He has one book—the Bible which his grandfather bought for his grandmother years ago when they had a good crop. He is proud of it, and keeps it in a place of honor on the shelf over the fireplace, but he can't read it. Illiteracy shuts him off from everything like a wall.

The effects of farm tenancy are felt by the whole state, not by tenants alone. Owners of tenant farms do not necessarily profit by this kind of land use. Some of them may manage to make money no matter what happens to anyone else. But when crops are poor, owners lose their shares just as the croppers do.

Every year more than a third of the state's tenant families pile their furniture into wagons or trucks and move to new farms.

Tobacco and cotton—the plants themselves and the way they are raised—both take a great deal out of the soil. Unless the land is taken care of, it becomes unable to produce a crop. And when land is not productive, everyone in the state suffers from the waste of resources.

The sharecropper's low income means a lower income for other people, because he is unable to buy many of the things they manufacture or sell. A sharecropper's low income means a lower income for the state, too. A man who owns little, earns little, and spends little is not expected to pay much in taxes. This is one of the reasons why North Carolina is not any richer in institutional wealth—so many people in the state have low incomes.

Here Are Some Questions to Think and Talk About

1. Why do not sharecroppers grow more food?

2. Why do they move so often?

3. Why are so many of them illiterate?

4. How does illiteracy help keep the sharecropper from improving his condition?

5. How does sharecropping as it exists in North Carolina
    a. Hurt land owners?
    b. Hurt the state?
    c. Help explain why North Carolina is poor in institutional wealth?

23. *Farm Owners*

JEFF BOWERS often says there's nothing like owning your own place. "What I want more than anything in this world," he tells Sarah, his wife, "is a home and a farm of my own." The man who rents always hopes that somehow, some day, this wish will come true.

This was not a wish, but a fact, for half the farmers of North Carolina in 1945. They owned their farms. These men discovered that the farm owner has worries the tenant never thought of. But they knew the independence and self-dependence of owning the soil they plowed and planted.

Farm owners in North Carolina do many kinds of farming.

1) Mort Crabtree lives in one of the mountain counties, in a gray cabin almost hidden by the trees which close in on all sides. Mort has no regular cash crop, and so little of his land is cleared that he can hardly be called a farmer. Yet he makes his living on the land, producing what he and Mary and the children need. He is a subsistence farmer.

The Crabtree family lives simply. Every spring Mort plants a vegetable garden. He keeps chickens, but they are not a special breed. It is just a mixed flock which

Some North Carolina farmers raise no cash crop, and grow only enough food for their own use.

supplies a few eggs and an old hen for dinner now and then. Pigs grunt in a pen beyond the fence. And now Mort is raising a calf because the teacher said the children ought to have fresh milk.

Four apple trees grow in the yard at the back. Although they need to be pruned and sprayed, there is enough fruit for Mary to make several kettlefuls of apple sauce every year.

Bees are buzzing around two hives near the corner of the fence. Most folks think there's nothing better than sourwood honey, but Mort likes sorghum on his bread, although he doesn't grow any cane for himself. He helps a neighbor with the cutting and grinding, Mary lends a hand with the boiling, and they carry home a few jugs of sorghum for their trouble.

In a field that tilts steeply against the sky, Mort raises corn for the family and the animals. He generally has enough left over to trade for coffee and salt and anything he doesn't produce for himself.

By the time the leaves are off the trees in the fall, Mary

has stored away Irish potatoes and sweet potatoes, and jars of pickled beans, sauerkraut, tomatoes, preserves, and fruits. After the weather turns cold, there'll be fresh sausage and spareribs for a while. A few smoked hams and shoulders and some sidemeat will hang in a safe place—meat for the winter, with squirrel stew once in a while when Mort feels like hunting.

If they need a little cash, Mort may be able to earn it by doing odd jobs. Or he may have to sell more food than they can spare. Mary sometimes manages to save a dozen eggs. The money helps to buy shoes for the children.

There are many subsistence farmers like Mort Crabtree in the Carolina Highlands and in the Piedmont, especially the central part. There are some along the coast. And there are others scattered here and there over the state. Their way of life varies from section to section, but they have one thing in common: they are almost self-sufficient.

The individual subsistence farmer produces little more than he and his family need for a simple living. He has no big cash crop. Farming is not a business with him.

2) Jim Nash is not like Mort Crabtree, growing almost everything he needs. Jim buys almost everything he needs with the money from what he grows. He is a commercial, cash-crop farmer.

He lives in one of the counties of the Coastal Plain, where the country is flat and open. Cotton used to be his money crop, but now most of his income comes from tobacco.

Jim superintends the production of the crop as if he were a foreman in a factory, superintending production.

This keeps him busy. He doesn't do any farm work himself. The land which he farms is worked by hired men.

Jim and his manager plan the work for these men and tell them what to do. He owns all the tools, machinery, and work animals; he buys the seed and the fertilizer; he pays the hired men their wages. The entire expense of producing a crop comes out of his pocket, and the profit—or the loss—is his.

Part of his land is divided into a dozen small farms which are worked by sharecroppers. Jim spends some of his time riding around to see how his tenants are getting along, advising them about their crops, and often telling them what they ought to do. He wants them to have good crops. He receives half of each crop for rent, and the tenants pay him out of their own shares for the food and fertilizer he has furnished.

Jim owns the store where the sharecroppers get their meat and other supplies, but he doesn't let the storekeeper overcharge for things, as some men do. "I want every one of my tenants to keep a record of what he gets," Jim says, "just as we do at the store. Then when it's time to settle, we don't have any trouble." Even so, Jim generally makes money at the store.

Since Jim buys seed and fertilizer for the croppers as well as for himself, he is able to order in large quantities; and he pays prices which are lower than a farmer pays when he is buying enough for only a few acres. Wholesale buying reduces the cost of producing a crop, and makes the profit larger.

Jim keeps chickens and a couple of cows on the place, but he has to buy most of their feed. He doesn't grow enough for the work animals and other livestock he owns.

Behind the house, out beyond the roses, one of Jim's

hired men plants a garden where he grows enough vegetables for the Nash family to use during the summer. But Margaret Nash never cans anything. In the winter she drives to the near-by town three or four times a week and buys fresh vegetables at the grocer's. The shelves in her storeroom are stacked with cans and jars, but they were filled in commercial canneries and not in the kitchen.

Life for the Nash family is not very different from what it would be if they lived in town. In fact, they often talk about moving in. The children go there to school; most of the family shopping is done there; on Sundays they drive to the church on West Main Street; Jim is a member of the board of directors of the First National Bank; Margaret belongs to the study club and the garden club. Many of their friends live in town, including some of the families from neighboring farms who left managers living on the home place. Every morning the owners drive out to their offices on the farms just as other men go to their offices on Main Street.

Farming is not a way of life for Jim Nash. It is a business.

Some North Carolina farmers have a great deal of land. They farm part of it themselves, with the help of men working by the day. Part of it they divide into small farms where sharecroppers raise the crops.

3) Dan Parker has a seventy-five-acre farm in one of the cotton-belt counties stretching along the South Carolina border. Like Jim Nash, Dan has a cash crop. And like Mort Crabtree, he grows as much as he can of what he needs. Dan combines some of the characteristics of the subsistence farmer and of the commercial, cash-crop farmer. He is a live-at-home farmer.

His cash crop is cotton, and he is just as careful as Jim is to buy the best seed and the right kind of fertilizer, because he wants to have a good crop. But he doesn't use all his cultivated land for a cash crop, the way Jim Nash does. Dan plants only fifteen of his acres in cotton.

"That's all I have time to take care of," he says, "and it generally brings in enough cash to get along on. I have other things to keep me busy."

He spends a lot of his time producing what he and his family need. "Why should we buy anything we can grow," he asks, "when I can raise it better and cheaper myself?"

So Dan and Nellie and the children live at home, without going to the butcher, the baker, the grocer, or the milkman for what they eat. The profit from the cash crop is money in the bank.

Dan keeps three cows—Jerseys famous among the neighbors for their milk. He raises plenty of hogs. His chickens are all Barred Rocks, not common mixed breeds like Mort Crabtree's.

"It doesn't pay to have no 'count chickens," Dan says. "They eat just as much feed as if they were good chickens. But they don't lay as many eggs, and they aren't as good to eat."

Dan raises turkeys and ducks, too. "What's the use of living on the land," he asks, "if we don't have some of the fat of the land that folks are always talking about?"

He grows all the grain and hay for his work animals and livestock, and he has regular pastures for grazing.

His vegetable garden isn't something he leaves to Nellie and the children, helping with it himself only when he hasn't anything else to do. Raising vegetables for the family is part of his job.

Some North Carolina farmers believe that raising food for the family is part of their job. They grow what they need themselves, and they often plan to raise extra potatoes or other vegetables as a cash crop.

It is Nellie's job to make sure that they will have plenty to last through the winter. She works at this all during the summer, with the two older girls helping. They can vegetables, berries, and fruits. They make preserves, jams, jellies, and pickles. The pressure cooker and the canning kettle are kept busy all summer long.

By winter, the storeroom shelves are crowded with rows and rows of jars and cans filled with food in such variety as Mary Crabtree never dreamed of—peaches in syrup, peach preserves, pickled peaches, apple sauce, pears, figs, plums, blackberries (the children gathered them by the bucketsful along the edges of the roads and fields), huckleberries, strawberries, cherries, and damsons, beets, peppers, carrots, vegetables for soup, tomatoes, corn, peas, beans, squash, relish, cucumber pickle, and watermelon rind pickle. Lard cans are full of dried peas and beans. Strings of red peppers hang on the wall.

Irish potatoes are piled on the floor of the smokehouse. And the meat from seven hogs is stored there, along with long strings of sausage, and cans full of lard.

Near the smokehouse door are five sweet potato hills, with enough potatoes to last until spring, maybe some to sell, and plenty for the plant bed, too.

Nellie Parker has a large house and family to take care of, and she does it well. To begin with, she is a good planner, and planning saves her both time and work. Besides that, she has all the important modern conveniences to help her with her job. She and Dan were among the first to get electricity out their way, and Nellie has electric helpers for almost everything.

Nellie does many other things besides housework. She helps the home demonstration agent at their group meetings. She is on the advisory board of the county library. She belongs to the Parent-Teacher Association. And she is always busy with some church work. "The more you do, the more you can do," Nellie says.

4) Duncan Anderson's farm is in one of the central Piedmont counties, not far from a branch of the Yadkin River. "I went off and worked in a mill for six months about thirty years ago," he says, "but I didn't like it. I was meant to be a farmer. I came back home, and I've been here ever since."

Some North Carolina farmers raise beef cattle or keep dairy herds. They think it is the best way for them to use some of their land.

Like Dan Parker, Duncan is a live-at-home farmer, but instead of having only one cash crop, he has several. He does diversified farming.

His father used to plant a lot of corn, even on land which sloped so sharply that rain washed the soil away. "He plowed his rows straight, too," Duncan says. "Right up and down the hills. Folks weren't doing much contour plowing in those days. The county agent told me it's a wonder we've got any farm left, the way the land's eroded."

Duncan uses the sloping land differently. He has made terraces to control erosion, and cover crops now hold the soil in place. Many of the hills are planted with permanent pastures for the dairy herd which produces one of Duncan's cash "crops"—milk and cream.

"It shows how times have changed," Duncan says. "My father couldn't have made money out of a herd even if he'd tried. In his day there wasn't any chance to sell dairy products. But since then, the state's built good roads, we've got motor trucks to haul things in, and when the mill expanded, a lot more folks moved to town to work. I can sell all the milk and cream I've got. And I haven't had any more trouble with erosion, either."

Duncan harvests some of his clover, lespedeza, and alfalfa for hay, keeps what he needs for livestock on the place, and sells the rest as a cash crop. He raises wheat as another cash crop, and he still plants corn on some of his land.

"Having a lot of livestock to look after, I've got something to do every day in the year," Duncan says. "It isn't like cotton or tobacco. They keep a man working every minute during part of the year, but the rest of the time he hasn't anything to do. He just lies around and does nothing."

Besides his diversified cash crops, Duncan raises the meats and vegetables to feed his family. He doesn't waste any time lying around.

The Andersons, like the Parkers, live at home.

These are some of the ways in which men who own their farms live on the land. There are other kinds of farming and different crops, depending on the size and location of the farm and the farming traditions inherited by the farmer.

Some farm owners in North Carolina are no better off than tenants. In fact, some tenants live better than some owners. Some owners are wealthy. And many make a comfortable living. All of them would probably agree with one of Dan Parker's neighbors, Clint Alderman. Clint says, "Farming's pretty work. I love it. The happiest man in the world is the farmer who can stand at his gate and know that he owns the land as far as he can see."

You'll Want to Add These Terms to Your Word List

subsistence farming
commercial, cash crop farming
live-at-home farming
diversified farming

You'll Want to Think About and Discuss These Questions

1. What worries does a farm owner have which never bother tenants?

2. What is the difference between farming as a way of life and farming as a business?

3. Which of the different kinds of farming described in this chapter requires the most regular work? Which is likely to make best use of soil? To produce the steadiest income?

4. Which calls for the greatest technological skill? The most capital wealth?

5. Which is most commonly found in your county?

You'll Want to Show in Pictures or Drawings

What life and work are like on each kind of farm described.

You'll Want to Read

Chapman, *Successful Farming in the South,* Unit III.
Marty and Company, by Knox, an easy-to-read book about life on a Carolina farm.

24. *North Carolina Needs More Livestock*

THERE is a great deal of unused land in the state. Some of it will produce an income, instead of lying idle, if it is turned into pastures for livestock.

North Carolina has land which is being misused, too—acres and acres which have been planted with cultivated row crops until the soil is exhausted or washed away. Many of these acres will yield a better income if the farmers will plant crops which cover the land. Cover crops hold the soil in place and give it back some of its lost fertility. At the same time they supply feed for livestock, and livestock may mean more money for the farmers.

Animal production—or animal husbandry, as it is often called—makes it possible for the farmers to use their time, as well as their land, to better advantage. When a man has chickens or cows or hogs or cattle to look after, he has something to do day in and day out through the whole year. During the winter months when a field-crop farmer is often idle, the man with livestock is earning money. The animals he looks after are a source of future income.

For years, North Carolina has failed to raise enough livestock. And for years, North Carolina has failed to grow enough feed even for the small number of animals in the state. Farmers have been buying most of their feed.

To make livestock profitable, farmers must start growing their feed. If they have cattle, sheep, or dairy herds, they must have year-round pastures wherever the winters are mild. In the colder parts of the state, they must grow more hays and feed for winter use.

Some time ago it would have been difficult to grow pastures for grazing the year round. Then men interested in animal husbandry started experimenting with various clovers, grasses, and other cover crops. They were trying to find out what will grow best in North Carolina and will make good pastures in different seasons.

Hugh MacRae of Wilmington spent forty years experimenting with crops for twelve-month pastures. He had to find out what plants would stay green during the winter, what plants would live through hot summers and droughts, which would grow back in a short time after grazing, and which would be permanent, growing back year after year with only one planting. With the pastures suggested by Hugh MacRae, farmers on the coast can have year-round grazing for their cattle.

Farmers in some parts of North Carolina have green pastures twelve months of the year. This dairy herd in Pender County can graze all through the winter.

Specialists at the North Carolina Experiment Station have carried out many tests, trying to find the plants which are best for pastures or hay in the different sections of the state. And men of the Tennessee Valley Authority have been doing the same thing for the counties in their area.

Studying the results of all these experiments, North Carolina farmers may now choose the plants which seem best suited to their particular soil and climate.

In the hottest part of the summer, especially when it is dry, North Carolina pastures may not be as good as those in other states. But during winter months, when livestock in the North has to be sheltered and fed in barns, animals are grazing in many parts of North Carolina. The milder climate saves both money and work for the North Carolina farmer.

Whether a farmer has a large commercial dairy herd or only one cow to supply his family with milk, he ought to have a breed or a cross-breed which is known to give quantities of good quality milk. When he has beef cattle, he should raise animals which have been proved to produce meat of tenderness and flavor. If he is raising chickens to have eggs, he should buy the kinds which are good layers. If he is raising them to sell as fryers or hens, he should buy others which are best for meat. His hogs and sheep should be tested for the quality of their meat and their resistance to disease.

Through the state's technological and institutional wealth, information about livestock is available to North Carolina farmers. Men at the Agricultural Experiment Station in Raleigh carry on experiments and tests, and they study the reports of what has been done in other parts of the country. As a result of this work, county agents are able to advise farmers about the kinds of livestock which have been found best for various purposes.

North Carolina can produce more livestock, and North Carolina needs more livestock. The farmers do not have enough for themselves. According to the 1945 Census of Agriculture, about fifteen farmers out of every hundred reported no chickens or other kind of poultry. About thirty-nine out of every hundred reported no milk cows, and about thirty-three reported no hogs or pigs.

Some farm families without livestock of their own buy their milk and meats, chickens and eggs. If these things were raised at home, the farmers would have to spend less for food, and the money saved could be used for something else.

But many of the farm families without livestock are families which haven't enough money to buy such expensive foods. So what happens? They simply do without.

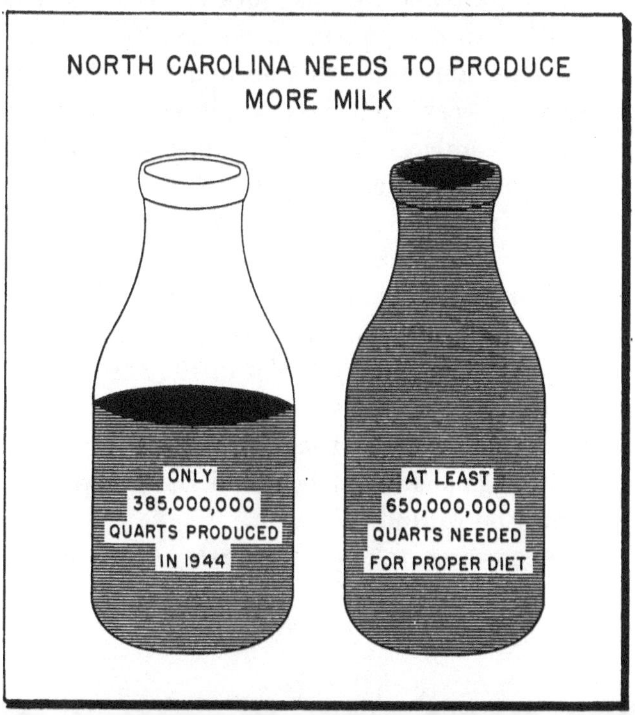

These families need to keep cows and chickens and pigs in order to get some of the foods which are an important part of the daily diet.

There are many other consumers in the state—people living in our towns and cities who produce little or nothing for themselves. They must buy almost everything they have to eat. Yet much of the food that is sold over the counters of North Carolina stores has to be brought in from outside the state, because North Carolina farmers are not growing enough to supply these markets.

The State Department of Conservation and Development has collected information about these markets for food within North Carolina. In 1941, North Carolinians ate about 189 million pounds of beef, but only 36 million pounds came from North Carolina farms. The rest was shipped in.

What about eggs and poultry? The 1940 census gave us the answer to this question. If Tar Heels had been limited to the poultry produced by farmers in the state, each person in North Carolina could have had only three eggs a week that year, and there would have been a chicken apiece about five times a year. The rest of the eggs and chickens we ate in North Carolina that year came from other states, or were raised for themselves by people who were not farmers.

North Carolina is especially well suited to raising chickens and eggs. And the state has made some progress in poultry production. Yet in 1946 there were many weeks when the North Carolina housewife went to do her marketing and found that there were no eggs, or that the only eggs she could buy had been shipped into the state from the Middle West.

According to the Department of Conservation and

Development, the average consumption of dairy products in North Carolina is very low—less than half the average for the whole country. Yet in 1941 farmers in the state did not produce even nearly enough to fill this need. Our dairy farmers have a large market right here in North Carolina if they supply what we are now buying from other states plus what we will buy when our average consumption is raised to what it ought to be.

North Carolina farmers have other markets in the North. Not far away by road or railroad are cities like Washington, Philadelphia, and New York, where millions of people depend for their food on what is shipped from other parts of the country. There is no reason why North Carolina farmers could not sell meat, poultry, eggs, and dairy products on these northern markets.

Land and man and the state will be better off when plant production and animal production in North Carolina are developed into a balanced agriculture.

It Will Be Helpful

1. To study again the table on page 137 to see how much income North Carolina received from livestock in 1944.

2. To read about animal production. See Chapman, *Successful Farming in the South*, Unit IV, and Hammonds and Woods, *Today's Agriculture*, Chapter 26.

3. To check up on your understanding of these terms: consumer, consumption, animal husbandry.

4. To represent in a picture graph the comparison between the amount of beef, eggs, and poultry produced in North Carolina and the amount consumed.

5. To list the ways in which man, land, and the state will be benefited when animal production is greatly increased.

6. To learn how your own county ranks in the production of eggs, poultry, butter, milk, etc. Your farm agent can help you find this information. Or you may find it for yourself in *Agricultural Statistics*, published by the State Department of Agriculture.

7. To have a report about the way 4-H Clubs, Future Farmers of America, and New Farmers of America help increase animal production.

8. To write an article suitable for publication in a magazine for young people in which you point out the need for increased animal production in North Carolina. Emphasize the importance of quality as well as quantity, and show some of the things necessary to get started toward better and greater production.

25. *North Carolina Needs More Diversified Farming*

NORTH CAROLINA has a variety of minor crops, one of the greatest varieties produced in any state. Yet the value of these minor crops as sources of farm income is small. A lot of them are grown for home use and not for sale. Many farmers continue to plant only tobacco or cotton, even when they could and should grow other things along with these important cash crops.

Soils and climate in North Carolina make it possible for farmers to raise a variety of crops and livestock. Man—and not the land—is responsible for North Carolina's emphasis on non-food, cash-crop agriculture.

Farming experts of the North Carolina Agricultural Extension Service think that diversified farming will mean better use of the land and a better living for farmers. After making a careful study of the possibilities in the state, they have suggested changes suited to the various areas.

North Carolina is divided into three great geographic regions. There is the Coastal Plain, containing the Tidewater and the Upper Coastal Plain. There is the Piedmont. And there is the Mountain region.

North Carolina may also be divided into agricultural regions. Crop areas, they are often called. There are eleven of these large crop areas in the state. The map on page 191 shows their location.

Each area has, in general, the same kind of soils and climate and grows the same kind of crops.

Area 1 is the Northern Tidewater. The land is often swampy—so low and flat that ditches have to be dug for the water to drain off the fields. The main crops are corn and soybeans. The minor crops are cotton, Irish potatoes, sweet potatoes, and various truck crops. Much of the corn and some of the soybeans are used to feed hogs.

Better drainage is needed on land that is used for field crops, and most of the less productive land should be made into pastures for raising beef cattle on a commercial basis. Lespedeza, velvet beans, and vetch could be planted, as well as such fruits as dewberries, blackberries, strawberries, peaches, and watermelons.

Area 2 is the Southern Tidewater. Although the land is less swampy than in Area 1, much of it drains slowly. Fertile, productive soil is often lying idle because it is too wet. And land unfit for use is in cultivation. Some way should be found to help farmers move from poor land to land that is fertile.

The major crops are corn, tobacco, and cotton. The minor crops are soybeans, peanuts, sweet potatoes, Irish potatoes, berries, other truck crops (such as beans, cucumbers, and lettuce), and hogs.

This area should reduce the acreage in corn and cotton and increase the acreage of pastures. Dairy cattle, beef cattle, and poultry are suggested as money crops.

Area 3 is the North Central Coastal Plain. The land here is gently rolling, and most of it is well drained. Corn, cotton, and peanuts are the main field crops. Some tobacco is raised.

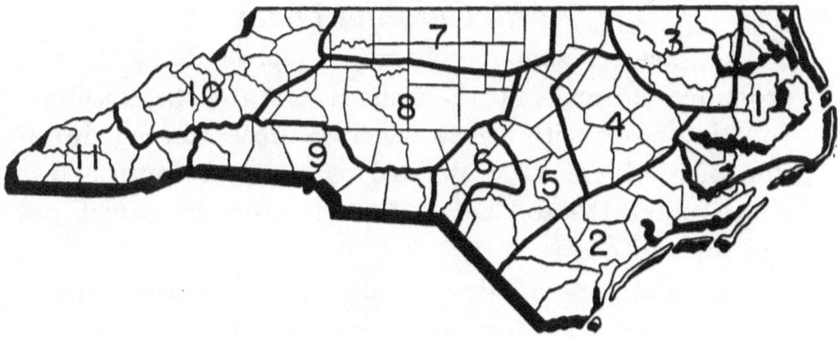

CROP AREAS IN NORTH CAROLINA

Soil improvement is needed more than anything else. For a long time field crops grown without much fertilizer have been using up plant foods and organic matter. The fertility of the soil must be restored even if this means fewer acres in peanuts, corn, and cotton.

The land which is being improved will still produce an income if the soil-improvement crops are used in feeding livestock. Farmers in this area already raise many hogs, which are fattened on corn and waste peanuts. More hogs, dairy herds, beef cattle, and poultry would make good use of cover crops.

Area 4 is the Central Coastal Plain region. The land rolls gently and is well drained. Cotton and corn are grown here, but the major crop is tobacco. This is the heart of the bright-leaf, flue-cured tobacco belt of North Carolina, one of the greatest tobacco-producing regions in the world.

For a long time it has also been the leading area for farm tenancy in the state, with a high percentage of sharecroppers. Commercial fertilizer is used in large quantities. Very little attention is paid to improving the soil.

The farmers should continue to grow tobacco, and probably cotton, but they should rotate these soil-exhausting crops with others which improve the soil. Rotation of crops must be planned carefully, because plants which make the soil better for one crop are sometimes bad for another.

Hogs and some poultry are already being raised in this area, but there are few milk cows. Livestock is needed—dairy herds, cattle, sheep, and more poultry—so that families will be producing meat, eggs, and dairy foods for their own use. And the farmers should also grow the feed for their livestock.

Area 5 is the Upper Coastal Plain region. Most of this area is in the Coastal Plain, where the land is nearly level. The rest is in the Piedmont, where the land is rolling. Drainage is good.

Cotton occupies a large acreage. Corn and tobacco are also important crops.

Here, too, as in Area 4, the soil has been neglected, and a great deal of commercial fertilizer is used. Erosion, and not drainage, is a serious problem, since farmers are growing cultivated row crops in rolling country.

Agricultural experts suggest that there should be fewer acres of row crops and more acres of soil-improvement cover crops. Winter cover crops, would keep the land from washing away. And if these crops were legumes, they would add valuable organic matter to the soil if they were plowed under in the spring.

This area also needs more livestock for home use, particularly dairy cattle and poultry. And many of the farmers need to grow more of their feed, instead of buying it as they have in the past.

Some of the minor crops which are well adapted to the

The soybean is one of the crops which is well suited to the soils and climate in some parts of the state.

various soils are blackberries, bunch grapes, cantaloupes, dewberries, peaches, barley, grasses, lespedeza, rye, early Irish potatoes, early and late sweet potatoes, and watermelons.

Area 6 is the Sandhills region of North Carolina. The soil is coarse, and since the land is quite open, water runs off or soaks in very quickly. There is danger of leaching and, in some sections, excessive erosion.

Cotton is the main crop. The minor crops are corn, tobacco, small grains, peaches, dewberries, and blackberries.

Soil improvement is needed so badly that many acres of cotton should be replaced by crops like velvet beans and vetch. The farmers should keep more dairy cattle and probably more livestock in general, as there are few animals in this area. There are other minor crops well suited to this section: bunch grapes, cantaloupes, early sweet potatoes, velvet beans, and watermelons.

Area 7 is the Northern Piedmont. The profile of the land rises from the Piedmont hills to the mountains in the west. Damage from erosion in this area is severe.

Tobacco is the main crop. A large acreage is planted with corn, which is grown as feed for the mules and horses. Both corn and work animals are here because they are needed for the tobacco crop.

This is the Old Tobacco Belt. The average farmer often keeps the soil poor deliberately, and depends on commercial fertilizers to make a crop. In this way he can use the mixture of plant foods which will produce the kind of tobacco he wants. He likes to let a field stand idle for a year or two before he plants it with tobacco again. Since the land slopes sharply, erosion has made many of these idle, unprotected fields worthless either for tobacco or for anything else.

This area needs terracing to save the soil, and a rotation of crops to improve its natural fertility. Farmers should increase the acreage of lespedeza, clovers, soybeans, and cowpeas, and perhaps develop other minor crops like apples, peaches, vegetables, barley, rye, and wheat.

Area 8 is the Central Piedmont. Most of the land is rolling, with low mountains in some of the counties and higher mountains in the west. Erosion is common.

In some sections of North Carolina, men raise flowers instead of tobacco or cotton. These workers are harvesting one of the flower crops —daffodils.

Wheat, corn, and cotton are the major crops. The minor crops are tobacco, oats, barley, rye, hay, and fruits. This is the main area in the state for red clover, sweet clover, lespedeza, and alfalfa.

Many of the farmers own their land. More livestock is raised here and farming is more diversified than anywhere else in North Carolina. Grains and hay are grown for livestock, dairies are numerous, and the farmers generally raise enough hogs for their own use and enough poultry for themselves and the local markets.

Farmers in this area are already improving their soil, but much of the land still needs terracing and still needs to have legume crops plowed under. And there is land lying idle which could be put to good use in raising more beef cattle and more sheep.

Area 9 is the Lower Piedmont. The land is rolling or hilly, and there is good drainage—much too rapid for the clean-culture, row crops raised by most of the farmers.

The major crop is cotton, and the other important crop is corn. The two together occupy over three-fourths of the land in cultivation. The combination of row crops and hilly country has allowed this area to be badly cut by erosion.

Practically all the land which is used for cultivated crops needs to be terraced, or planted in a way that will keep the soil in place. On much of the land, cultivated crops should be replaced by small grains, and legumes like lespedeza, soybeans, cowpeas, vetch, and crimson clover.

There is already more livestock here than in the Coastal Plain. Farmers have dairy herds because they are able to sell milk and butter and other products to the people in the manufacturing towns of the Piedmont.

Minor crops which might be increased are apples, barley, cowpeas, grasses, lespedeza, peaches, rye, and soybeans.

Area 10 is the Northern Mountain region. This is mountain country. Much of the land now in use is so steep that it should never have been cleared. And once cleared, it should never have been planted with cultivated row crops.

Yet corn is one of the main crops in this area. Small grain and hay are also raised, as well as apples, Irish potatoes, cabbage, and burley tobacco.

Since grasses grow well in this part of North Carolina, the pastures are good. There are more beef cattle, sheep, and poultry here than in any other area of the state, and the farmers also raise enough hogs for their own use.

Pastures should be increased by reducing the corn acreage and sowing the steep cultivated fields in grasses, small grains, lespedeza, and clovers. This change would help to control erosion. And if livestock is raised on these pastures, the crops which save and improve the soil will also make money for the farmers.

Area 11 is the Southern Mountain region. The profile of the land is much like that in Area 10, except that the valleys are larger and the mountains are not so high.

Farms are small. Many of the men are subsistence farmers who work at other jobs part of the time. Some of them are simply subsistence farmers who sell a little of their produce now and then.

Corn is the major crop. The minor crops are small grains, hay, fruits, and Irish potatoes. Livestock is plentiful, especially beef cattle.

Farmers in this area, like those in Area 10, are cultivating land which is steep. These sloping fields should

Sheep can' be raised in North Carolina both for their wool and for their meat.

be turned into pastures, and the number of livestock should be increased. More vegetables could be raised here for a cash income.

These are the main suggestions of agricultural leaders for a better farm program in North Carolina. Each farmer, after studying the program and talking things over with his county agent and perhaps his neighbors, must decide for himself how to fit the suggestions to his own land and markets.

Years ago it would have been difficult to sell some of these suggested food crops even if the farmers had raised them. In those days the state was more rural than it is today. There were not many towns or cities where food could be sold in any quantity. And since meats, fruits, and vegetables are perishable, they could not be shipped to distant markets without spoiling.

Then, as North Carolina developed, towns and cities grew larger, and there were more people depending on others to grow food for them. With modern transportation, fruits and vegetables picked one day may be eaten a day or two later anywhere in the state. Trucks take crops along state roads from one end of North Carolina to the other. Perishable foods, now kept cool by refrigeration, are hauled rapidly over a network of highways and railroads to markets in the North. Some go by air express.

These developments in markets and shipping were necessary before North Carolina farmers could sell a variety of food crops.

Review

1. Meaning of these terms: rotation, terracing, leaching, legumes, erosion, organic matter, row crops, cash crops, field crops, cover crops, soil improvement crops, soil exhausting crops.

2. Facts about the different sections of North Carolina and about North Carolina soils in Chapters 3 and 4.

Make

1. A large map of the area of the state in which your county is located. On one side list the chief facts about the condition of the land and about the crops grown. On the other, list the suggestions given for a better farm program.

2. A field trip to some of the farms in your county where improved farm practices are carried on. Ask your agriculture teacher or farm agent for help.

Suppose

The farmers in your area have decided to undertake a a program of diversified farming along the lines suggested in this chapter.

1. Where or from whom could they get advice and help?

2. What improvements in transportation and in marketing would be needed to make their diversified farming pay?

26. *Better Use of the Land*

THE early settlers in the state found "mild Winters and a fertile Soil beyond Expectation, producing everything that was planted to a prodigious Increase; their Cattle, Horses, Sheep, and Swine . . . passing the Winter without any assistance from the Planter; so that everything seemed to come by Nature, the Husbandman living almost void of Care, and free from those Fatigues which are absolutely requisite in Winter Countries, for providing Fodder and other Necessaries."

So wrote an Englishman, John Lawson, early in the 1700's.

"The Inhabitants of Carolina," he continued, "through the Richness of the Soil, live an easy and pleasant Life." Lawson found it "a Country that, with moderate Industry, will afford all the Necessaries of Life."

William Byrd, when he was helping to lay out the dividing line between Carolina and Virginia in 1728, noticed "the great felicity of the Climate, the easiness of raising Provisions, and the Slothfulness of the People." He found the food "extremely cheap, and extremely good, so that People may live plentifully at trifling expense."

"Surely," Byrd wrote, "there is no place in the World where the Inhabitants live with less Labour than in North Carolina."

Times have changed since those early years of the eighteenth century, and today people on farms do not always "live plentifully." Many farmers are using all their land and their labor to produce only cash crops—generally tobacco and cotton. Farms specialize in perhaps one product, as factories often do, and the farm families depend on cash from the specialized crops to buy everything they need.

But sometimes there is very little money. And since food is no longer "extremely cheap," the farmers who do not raise their own food and the feed for their livestock find that they and their families are doing without many of the things they want, often without some of the things they should have.

This is why in North Carolina there are people living on the land without many of the things the land produces, sometimes not even having enough to eat.

North Carolina farmers will have more of a chance to know the "easy and pleasant Life" that John Lawson saw here when they are once more doing what the early settlers did—producing for themselves "the Necessaries of Life." When they do this, they will be truly living on the land.

"Look at all the animals Uncle Bill has. He's going in for livestock in a big way—in addition to cotton and tobacco."

There is nothing new about North Carolina's need for live-at-home farming. People have been talking and writing about it for a long time.

In 1870 a North Carolinian wrote to a state farm journal called *The Reconstructed Farmer*, "Many of our farmers . . . buy meat, corn, seed and manures and every necessary article of family consumption from the North."

Less than twenty years later *The Progressive Farmer* declared, "Hard times will hover around our firesides so long as we buy meat and bread, hay and fertilizers, and other farm supplies, and attempt to pay for them from the proceeds of one crop."

In 1920 North Carolina was producing little more than half the feed and food used in the state. In that year it was estimated that the bill for "imported" food and feed was $235,000,000.

In 1922 Governor Cameron Morrison urged the people to grow more of their own food, saying, "Our whole agricultural life is weakened by the fact that we have neglected the production of sufficient food on the farms of the State to make our agricultural life independent."

"*Isn't Dad* wonderful? *He grows* every*thing—not just tobacco.*"

In 1929 the state was still importing much of its food and feed, probably more than in 1920. And the value of the cotton and tobacco crops was so low that farmers hardly had enough money to buy anything to eat.

Once more the state's governor urged the people to grow what they needed. In January, 1930, Governor O. Max Gardner began a statewide campaign with the slogan *North Carolina Must Live-At-Home.* "The live-at-home idea," he said, "increases the ability of our own folks not only to *have without buying* most of the goods for which they spend the most money, but . . . to demand for their cash crops better prices than they have ever been able to do up to now."

During the Second World War it became more important than ever for the people in North Carolina to raise all the food they could—at least enough for themselves.

Live-at-home farming today does not mean returning to the simple, self-sufficient living of the early settlers. It means, instead, the use of modern methods and machines wherever they are practical, of tested seeds and tested livestock, of electricity and running water for

"*You see, Jim, I got this big check because I grew a* high grade *tobacco!*"

health and for convenience. It means raising the things that can be raised more cheaply than they can be bought.

No single set of rules can be made which will fit all the farms in North Carolina, because conditions are not the same everywhere. In the east, where the land is flat, water drains off very slowly. In many other parts of the state the land slopes so sharply that much of the soil washes away with the rains. Farming problems vary from one part of the state to another, often from one farm to the next, sometimes on one farm alone.

The program for this improvement of agriculture in North Carolina is a general one, planned by experts for better use of the land in the whole state. Every farmer will have to study this program and adapt it to his own particular circumstances and needs.

Agricultural leaders suggest that our farmers should

1) Raise more livestock—poultry, dairy herds, beef cattle, and other meat animals—which would make a better balance between plant production and animal production.

"Everything on this table today—the meat, the vegetables, the milk and the butter—we raised right on our own farm."

2) Develop diversified farming, growing more of the crops which North Carolina already produces in great variety but not in great quantity—fruits, peanuts, soybeans, potatoes, and other vegetables, as well as livestock and poultry.

3) Continue to plant tobacco and cotton for cash income, but only when the crops will be of high quality.

4) Practice self-sufficient, live-at-home farming—each farmer producing what he needs, as much as possible, and all farmers producing what the state needs, as much as possible.

5) Conserve and improve the soils—by rotating crops and by plowing under legumes and other humus crops.

6) Control erosion—by terracing; strip cropping; planting small grains or other cover crops; establishing pastures on land not suited to field crops; and reforesting land which is otherwise useless.

7) Drain wet fields—by digging ditches and putting in pipes and drains where the land is so flat that the water will not run off.

"What am I doing? I'm plowing under these beans to make my soil richer."

North Carolina soil has changed in the last three hundred and fifty years. It is no longer what Barlowe described in 1584 as "the most plentifull, sweete, fruitfull and wholsome of all the worlde." Our use of the land has wasted this natural wealth.

John Lawson noticed that farmers had good crops without much care or work. "If our Planters," he wrote, "when they found such great Increase, would be so curious as to make nice Observations of the Soil and other remarkable Accidents, they would soon be acquainted with the Nature of the Earth and Climate and be better qualified to manage their Agriculture to more Certainty and greater Advantage. For I must confess I never saw one Acre of Land managed as it ought to be in Carolina since I knew it."

There have always been a few men who tried to save their soil and improve it, but there have been many others who farmed in a way that exhausted and wasted the land. Today, after years of waste, farmers all over the state are beginning to "manage their Agriculture to more Certainty and greater Advantage."

"You see, Dickie, our land doesn't wash away like Uncle Jim's. Our trees and our new terraces and strips all help us keep our soil."

In the hilly part of North Carolina, erosion must be controlled or the land will wash away. Terracing, contour plowing, and strip cropping help to keep the soil in place. Small grains, winter legumes, and similar crops protect the land because they cover the fields instead of being planted in rows. Crops for pastures hold the soil, and so do the roots and fallen leaves of trees.

On the coast, fertile land now lying wet and useless must be drained if it is ever to become productive.

All over the state, farmers must rotate their crops, so that one crop returns to the soil some of the plant foods which the last crop has used and the next crop will need. Legumes and other soil-improvement crops must be plowed under, to increase the organic matter in the soil and add to its fertility.

This costs a lot of money, but it is necessary. And in the end it makes larger profits.

Many of these methods of saving and improving the soil had been followed in Europe long before the settlement of America, and they were known and used by individual planters in North Carolina and the other colonies.

"We'll open that ditch over there so the water will drain off."

Washington and Jefferson and the great agricultural leader, John Taylor, all worked out systems for rotating their crops. Taylor advised the use of green manures. By 1825, contour plowing on hills, hillside ditching, and terracing were accepted as ways to control erosion. These farming methods were known by a few, but they were not generally practiced.

Judge Thomas Ruffin of North Carolina, speaking before the State Agricultural Society in 1855, said: "The soil is not what it once was, and our task is not merely to preserve fertility but in a great degree to restore that which has been more or less exhausted."

This is still the job before North Carolina today.

You'll Remember More if You

1. List in your notebook the seven suggestions for better use of the land given in this chapter.

2. Prepare a large illustrated wall chart or series of smaller ones listing these suggestions.

You'll Learn More About Good Land Use if You

1. Look through and study some of the pamphlets and materials about conservation, farm practices, etc. which your agriculture teacher, farm agent, Soil Conservation representative, 4-H Club leaders and others have given or will give you. Also see Unit II in Chapman, *Successful Farming in the South.*

2. Arrange to see some of the slides or movies prepared by the U. S. Department of Agriculture and available through the State Extension Offices or the Bureau of Audio-Visual Aids at Chapel Hill. Some of these films are: *Live At Home, Food and Soil, For Years to Come,* and *The South Grows Green.*

27. *Working Together Toward Better Farming*

FARMERS the world over have always been looking for some way to make more money from their crops. Many of them used to do this simply by growing larger crops, until they found that they were sometimes producing more than they could sell at any price. Other farmers thought they could make money by growing better crops than anyone else.

To do this, however, a farmer needs more information than one man can easily get for himself. So through their government, farmers have set up agencies to study the problem of producing better crops and to give farmers advice and help.

Experts in agriculture knew the truth of what a few men had been saying for a long time—that the wealth of the land was being wasted by misuse. Many farmers knew this, too. They knew that thousands of men, making only a poor living on the soil, were unable to break away from the old systems of farming. But no one man, working alone, could do much about it.

Long ago people in North Carolina started working to make agriculture better. In 1877 the legislature established a State Department of Agriculture which now has many special divisions.

With money borrowed from the government many tenants have been able to buy land, build a house, and start out for themselves.

Experts in the Department make tests of fertilizer, feeds, insecticides, and seeds to make sure that they come up to certain standards. They try to find ways to wipe out destructive insects and livestock diseases. They check the standards of milk and butter fat produced by dairy herds in the state. They test soils to find out whether farmers need to add plant foods and other chemicals. They report on crops and livestock being produced in the state. They advise farmers about the grading and packing of products they have for sale. They supervise the credit unions through which farmers may borrow money, and they carry on other work helpful to farmers.

Many of the courses of study at the North Carolina State College of Agriculture and Engineering in Raleigh and at the Agricultural and Technical College in Greensboro are planned especially for students interested in general farming, dairying, livestock, poultry, fruits, vegetables, horticulture, or field crops. State College has Agricultural and Dairy Foundations which help to support special research.

In cooperation with the State Department of Agriculture, State College operates a number of test farms. These are really field laboratories where scientists try to find ways to grow better crops.

In cooperation with the United States Department of Agriculture, the Agricultural Extension Service of State College has a farm agent in every county in the state. This county agent is a man who knows the practical problems of farming and who is also trained in agricultural theory and farm management. County agents keep farmers informed about new and better ways of doing old jobs and about federal farm programs. The agents try to help every farmer find out what crops are best suited to his land, and where and when to market them.

Home demonstration agents show farm women how to plan and cook well-balanced meals, how to can and store food for winter, how to sell extra farm products, how to make life healthier and more comfortable for everyone on the farm. The Home Economics Foundation at the Woman's College of the University of North Carolina carries on a program to improve home management.

Agriculture teachers in many of our rural high schools are teaching boys the skills necessary to good farming. And through the 4-H Clubs, the Future Farmers of America, the New Farmers of America, and other clubs like them, North Carolina young people are learning ways to make farm living better.

The federal government also helps improve agriculture in North Carolina. For years the United States Department of Agriculture has been working to help farmers—testing and experimenting with seeds and livestock, diseases and pests; finding the best methods for growing, grading, and selling crops.

In the 1930's citizens of the nation, through the government, set up special agencies—among them the Farm Credit Administration, the Farm Security Administration, and the Agricultural Adjustment Agency. These agencies helped farmers all over the country toward better farming and better farm life.

In North Carolina, many tenants who had been hoping all their lives for a chance to own their farms were able to borrow money from the government with which to buy land, build a house, and start out for themselves. Farmers who had been trying to make a living on land

Poles holding electric wires and poles holding hay stacks often cross the same field in rural North Carolina.

which was poor and worthless for crops were helped to move to more fertile land.

Farmers needing livestock, seeds, tools, or other equipment could borrow enough money to pay for them. Farmers wanting to reduce their acreage in row crops like cotton were helped to make the change to soil-improvement crops.

The Rural Electrification Authority now makes it possible for farmers to have electricity to light their homes and run their machines. Farmers in many neighborhoods organized themselves into groups which helped to bring electric power to their communities.

The Soil Conservation Service encourages farmers to control erosion, grow crops which improve the soil, and plant trees or grasses on land that should not be used for anything else. Many of our farmers are taking part in this program. And with the help of the Soil Conservation Service, much progress is being made.

The Tennessee Valley Authority works with other agencies for better farming and soil conservation in some of the western counties. One aim of the TVA is to keep soil out of streams and reservoirs where it would reduce the flow of water. This is part of its program for controlling floods, making electric power, and keeping many of the rivers free for navigation.

Besides our government agencies, farmers have their own organizations for studying farm problems and suggesting policies and laws which help improve agriculture. There are the Grange, the Farm Bureau Federation, and the Farmers' Union. And there are other farm groups with special interests.

Members of the North Carolina Cotton Growers Cooperative Association are cotton farmers who work together in selling their crops. The Guernsey Breeders Association, the Jersey Cattle Club, the Swine Growers Association, the Hereford Association and similar groups are all trying to improve livestock in the state.

The Farmers Cooperative Exchange, with branches over the state, the Farmers' Federation of Western North Carolina, and independent mutual exchanges like the Farmers' Mutual Exchange in Durham all have markets where members may sell their products and buy what they need.

Farming and farm life in North Carolina must be made better, and this is a job for everyone in the state—farmers and non-farmers alike. Better use of the land is the first step in the people's struggle toward better living. This is the point at which the chain of things-as-they-are must be broken to begin a new chain of things-as-they-ought-to-be.

To do the job well, we must use everything we have—schools, colleges, and universities; experiment and re-

Men and women of a farm neighborhood meet together to talk about the work they plan to do in their community.

search in laboratories and on test farms; public and private funds; towns, cities, and counties; state and federal governments.

Then our natural resources—soils, moisture, and a long growing season—will produce wealth in abundance. And North Carolina will be nearer the technological, capital, and institutional wealth which rightfully belongs to the future of the state.

You'll Want To Read

Chapman, *Successful Farming in the South*, Units VI–IX.
Chapman, *Better Rural Careers.*
Stories about 4-H Clubs. The settings of these stories are not North Carolina, but the experiences of the boys and girls in them are typical of those in our state.
Allen, *Dynamo Farm.*
Lide, *Johnny of the 4-H Club.*
Porter, *Sandra Kendall of the 4-H Club.*

Now Is the Time!

To Look Backward

1. Is your word list up to date?

2. What are the chief ways to measure success in farming? In what ways does farming in North Carolina rank low? In what ways high? What are some of the things which account for her high rank? Her low rank?

3. How does your county rank according to different yardsticks? Do you know what steps should be taken to raise its rank?

4. How do agricultural experts or specialists learn how to advise farmers and what means do they use to make their ideas known?

5. Do you know what agency or agencies to go to for advice and information in case you want to:

a. Start a dairy?

b. Have your soil tested?

c. Learn what kind of wheat or corn is best adapted to your land?

d. Take a short course in fruit growing?

e. Do something about the gullies in your pastures?

f. Learn where to market your cotton?

g. Learn what kind of grass to plant in the yard of your home or school?

6. Imagine you are an agriculture teacher. Write an article for a newspaper or farm magazine describing the improvements in agriculture and in living conditions on the farms which have taken place in North Carolina in the last ten years.

To Look Around You More Closely

Observe the crops, land treatment, and farm activities in or near your community. Do you understand the "why" of these better than you did before? Do you understand how these activities are important to all citizens whether they are farmers or not? Are there signs of diversified farming?

To Look Ahead

See what comes next in your study about North Carolina. What agricultural products you have studied about are already important in North Carolina manufacturing? What others should be important?

Peanuts are already one of North Carolina's valuable crops.

IV. CREATING WEALTH BY MANUFACTURING

28. *The World We Live In*
29. *The Beginning of Manufacturing*
30. *Resources for Industry*
31. *Delivering the Goods*
32. *Power and Industry*
33. *Textiles*
34. *Tobacco*
35. *Forest Products*
36. *Other North Carolina Industries*
37. *What Factory Jobs Mean to Workers*
38. *People at Work*
39. *Diversified Manufacturing*

With skillful fingers as his chief tools, this worker is giving form to a piece of clay.

28. *The World We Live In*

THE machine age we live in today was a long time in the making. Man's first tools were his hands, plus the various natural "tools" he found lying around on the ground, such as sticks, rocks, shells, and the bones of animals.

After a while someone started improving these natural tools to make them more helpful. Perhaps he rubbed the edges of a rock to sharpen it, or tied it to a stick with a vine or a piece of animal skin to make it easier to handle. And from that time on, man began to search for ways of making tools which would help him do things better or faster or more easily—a search still going on today.

One man, working alone, could use the early tools by himself. But after the invention of more complicated tools, he needed more hands and more strength than he had by himself. He had to have a helper, sometimes two or three, to turn a wheel for power or to hold something in place while he worked on it.

As tools became heavier, there had to be something stronger than a man to operate them. Then animals or water were used for power.

Men gradually invented many kinds of tools, and began to combine them into machines. One man with a machine operated by power could do the work of several men.

At first each man owned his own tools and worked with them at his home. If he produced more than he needed for himself and his family, he sold the surplus. Or sometimes he spent all his time making things to trade or sell.

Then a new plan for producing goods developed—one in which a worker did not own either the machine he used or the things he made. A man with money to invest might buy several looms and set them up in a room. He bought yarn, and paid other men wages to come and weave it into cloth. The cloth belonged to him, and he sold it to anyone who would buy. This was the beginning of factories.

In colonial times in North Carolina there were a few wealthy planters who could afford to have things made for them in New England or Europe. And some people owned slaves to make things for them. But most families had to produce for themselves whatever they needed. Manufacturing was a simple home industry—the making by hand of articles at home for home use.

The first permanent settlers in Carolina—although they plowed the land and planted crops—spent some of their time making the things they needed in living and working. They were manufacturers as well as farmers.

In those days a man had to be something of a jack-of-all-trades. He was his own shoemaker, making shoes for everyone in his family. Before he could make shoes, he probably had to raise an animal, kill it, and skin it. He had to clean the hide, tan it, and treat it to make it soft. Then he cut the leather to the proper shape and sewed it with handmade tools and thread.

He had to be his own carpenter, building his house and furniture and sometimes making his own tools. To make furniture, he first had to cut down trees, cut the trees into lengths, and peel off the bark. He cut the wood into pieces of proper size, shaped them, and fastened them together. He probably used wooden pegs, as nails were made by hand and were scarce and expensive.

A woman not only prepared and cooked all the family meals. She made everything for household use: bedding, clothing, stockings, soap, candles, even medicines. And to make a suit or a dress two hundred and fifty years ago required a great deal of time and work. It was a job that had to be done by hand from the very beginning.

At first, wool or linen was used. The wool was sheared from sheep that had been raised on the place. Linen was made from flax that had to be planted and harvested and especially prepared before it could be spun. But woolen

There are still men in North Carolina making furniture by hand, like this worker who is shaping a chair rocker with a hand tool.

cloth was hot for North Carolina summers. And it took a lot of time and work to turn flax into yarn. Cotton became popular.

Whether wool or flax or cotton was used, it first had to be carded—or brushed—until the fibers were straight. Then the carded fibers were spun—joined closely together into one long strand of yarn. The yarn was woven into cloth on the family loom. Then came the cutting and sewing. There were no paper patterns; pins and needles were scarce and expensive; and thread had to be spun especially fine and strong for sewing. Everything was done by hand.

Any "spare" time a woman had during the day was spent at her spinning wheel, which always stood ready in a corner by the fireplace. The children helped—the young ones separating cotton lint from the seed, the older girls carding and spinning. In the early evening before bedtime, fingers kept busy by candlelight. It took constant work to supply a family with clothing.

The life we know today is very different from the life of the pioneers. Almost anything we need may be bought at stores ready-made. And many of the everyday chores which people used to do for themselves are now done mechanically or by men specializing in different kinds of work. It is no longer a handmade world.

A man seldom builds with his own hands the house he lives in. There are carpenters, bricklayers, all kinds of special workers who build most of the houses today.

Instead of open fireplaces or stoves in each room, a house may be heated by coal or gas or oil. And this fuel may be burned in a furnace which is operated mechanically so that the house is always warm. Water no longer

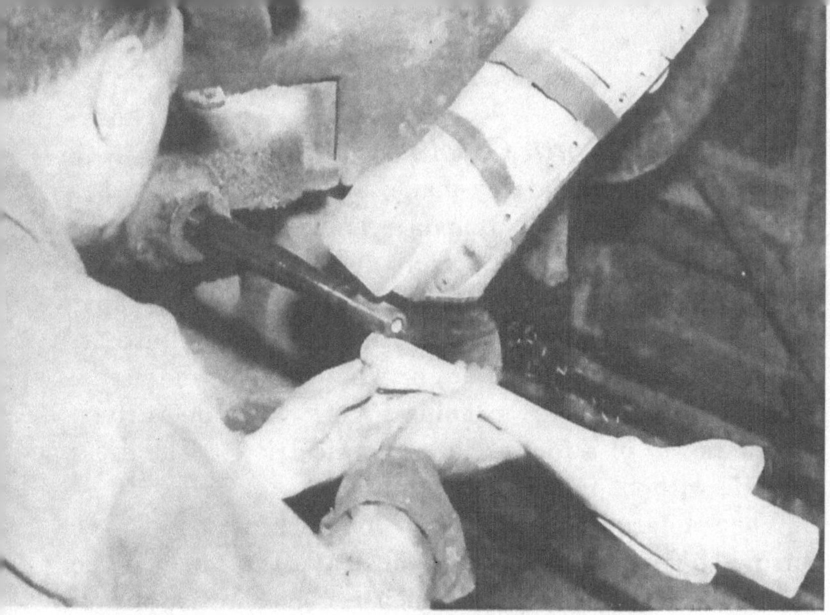

Most of our furniture is made in factories where much of the work is done with machines. Here a man is using a machine to shape the leg of a chair.

has to be carried into the house by the bucketful from a spring or a well or a pump. It may be piped in and ready for use whenever a faucet is turned on.

The stores are full of furniture and curtains and rugs. Instead of using candles or oil lamps or gas lights, people may flick a switch and have electric lights to make night seem brighter than day.

Grocery stores and markets are stocked with meats and with fruits and vegetables from anywhere in this country, from Hawaii or the Philippines or Cuba or Mexico, from South America or China or Europe. Food in cans and jars and packages stands ready on pantry shelves; or it may be kept in a mechanical refrigerator. With a can opener and modern cooking equipment a woman can prepare a complete meal with all the "fixin's" in less time than it used to take to get a fire started and heat a kettle of water two or three generations ago.

Shoes today are factory-made, and so are the clothes most people wear. If a woman feels like making a dress, bolts of material in many colors and fabrics are stacked on the department-store shelves. She can buy a paper dress-pattern; needles, pins, and spools of thread; buttons, snaps, or zippers; lace, braid, or whatever else she needs. And she has a sewing machine to sew with, one driven by the power of a foot pedal or by electricity. The dress is made without much time or work.

When a farmer plows, he no longer has to walk the furrows behind his mule. He can leave the mule in the stall and plow with a tractor which uses gasoline for power. A gasoline engine or electricity may help the farmer with most of his heavy work. And while he does his barnyard chores, he may listen over his radio to the music of a fifty-piece orchestra or the jokes of a popular comedian.

These changes are concerned only with simple, everyday matters—a house to live in, food to eat, clothes to wear. In a thousand other ways this life of ours is different from that of the early settlers. Yet it was not very long ago that people lived in a handmade world. As the world measures time, the pioneers were living only yesterday or the day before. What happened to change things so quickly and completely?

The world itself changes slowly. The earth—now as always—has soil and water, rock and minerals, animals and trees and vegetable growth. The thing that has changed quickly is not the world, but the people in it. Men have found out how to use the materials which have always been here, waiting. Man's knowledge and technological skills have made a new world out of the old one.

Years ago, in parts of Pennsylvania and western Vir-

With a special machine this worker is carving ornamental parts for furniture. One man can do work which would need many men if the carving were done by hand.

ginia farmers often saw oil floating on their streams, and men used to dig wells into the ground to get salt. These salt wells sometimes filled up with oil and had to be abandoned. The oil made them useless. People did not know what to do with a resource which today produces power and wealth. The oil was a nuisance.

For years the trees that grew tall and thick in Carolina were in the way. They were a handicap. Men cut them down to get rid of them, to get at the land they covered. It is a different story now, a story in which a new chapter is being written and the end of which is not yet known. Man's increasing knowledge is finding unheard of uses for a resource which was for so long a nuisance — wood.

For years, long before white men pushed their way into the Piedmont, the waters of Carolina rivers were falling from the mountains toward the lower lands of the east and the west. Men built factories along these rivers, converting the energy of falling water into direct power to run their machines. They were creating wealth out of a resource which had been useless because men hadn't known what to do with it.

Then, after 1900, men turned this energy into hydroelectric power. Out of the same resources which had long been useless and which for a time had been used for direct power, they created greater power—greater wealth.

The men who used water as a source of direct power had no idea that they were wasting this resource. So far as they knew, they were making the most of it. So far as they *knew*. It is knowledge that makes the difference. Today men know how to put this resource to better use.

Man is always finding new uses for familiar resources, or ways to use something which has been considered worthless. Things thrown away yesterday may mean better homes, more and better food, easier travel, quicker transportation, and simpler communication tomorrow.

As knowledge and skills increase, industries develop and change. New ones spring up overnight, as quickly as mushrooms after a warm rain. But sometimes old industries keep going long after they should, when there are better and more profitable ways of using the raw materials they process.

For this reason it is a good idea for people in North Carolina to ask themselves a few questions about industry in the state. What resources do we have? And are these resources creating as much wealth as possible? What resources are not being used? What resources could be better used?

In order to think about these questions intelligently

we must first look back a bit over the last two hundred years to find out how manufacturing in the state started —to see how the handmade world became a machine-made one.

You'll Want To Read or Reread

1. Arnett, *The Story of North Carolina*, Chapter 12.

2. Newsome and Lefler, *The Growth of North Carolina*, pp. 85–107, 244–256.

3. Warren, *North Carolina Yesterday and Today*, pp. 64–69, 151–152.

You'll Enjoy

1. Collecting pictures, models, or actual objects which show old and new ways of manufacturing and old and new products.

2. Doing some things in the old-fashioned way:
   a. Weaving on hand loom.
   b. Dying articles with home-made dyes.
   c. Making a split bottom chair.

3. Having someone read aloud descriptions of earlier ways of doing things found in some of the books listed at the end of Chapter 11.

4. Reading parts of Eaton, *Handicrafts of the Southern Highlands.*

5. Seeing the short film, *Patterns of American Rural Art*, which shows in color examples of rural crafts. Order from the Bureau of Audio-Visual Aids, University of North Carolina.

29. *The Beginning of Manufacturing*

EARLY in the 1800's something happened which made life a little easier for Carolina women: the first cotton mills in the state were built.

Few of the early mills made cloth. At first they had only machines for spinning yarn. Later on, the mill owners put in looms for weaving.

A man hauled the cotton he had grown to a neighborhood mill, where it was carded and spun. He left part of the yarn with the mill owner to pay for the spinning. The rest he took home to the women of the family, who wove it into cloth. A lot of work still had to be done at home before anyone could have a new dress or a suit, but the work began with the weaving instead of with the carding and spinning.

By 1850 a number of cotton mills had been built, especially in the Piedmont, where there was rapidly falling water for power. And there were small mills to do other things besides making yarn and cloth—mills to grind corn and wheat, saw mills, iron works, shoe factories, leather and tobacco factories, mills to make hats and furniture, harness, wagons, and guns.

These early North Carolina mills—the cotton mills and the others—were local industries, using raw materials from the neighborhood and selling to people who lived

Before there were any mills, a woman often spent a great deal of her time at her spinning wheel, spinning thread. This picture shows lengths of fiber already carded and hanging on a knob, ready to be added to the thread.

near by. There were several reasons why factories in North Carolina were to remain local industries until the 1880's.

Perhaps the most important of these reasons was an invention. Back in the eighteenth century, machines for spinning and weaving had made it easy for mills in New England and old England to produce cotton cloth much more quickly than was possible when this had to be done by hand. The mill owners wanted raw cotton in large quantities for their machines—all they could get hold of.

The South could grow plenty of cotton, but cotton lint had to be separated from the seeds by hand, and that was slow work. Hands couldn't do it fast enough to keep up with the speed of the machines for spinning and weaving. And then, in 1793, Eli Whitney thought of a way to do the job quickly with a machine—the cotton gin.

After Whitney invented the gin, cotton was for years king in the South. Planters in North Carolina and other southern states grew as much cotton as possible, and for a long time they had markets waiting for their crops.

Many people in North Carolina and in other parts of the South made money from the thousands of acres of land they planted in cotton. Instead of investing this money in factories to spin yarn and weave cloth, they often used their capital wealth to buy more land on which to grow more cotton. King Cotton ruled the South, but it was cotton in the field and not cotton in the factory.

During the War between the States all this changed. With many of the men away in the army, farming could not go on as usual. Because of the war, southern cotton could no longer be sold to factories in the North. A northern blockade of southern ports made it difficult for ships to slip through and deliver cotton to the factories in Eng-

After a woman finished spinning, she then set up her loom and started to weave the thread into cloth. The weaver in this picture is holding the shuttle in her left hand, ready to pass it through the warp to the right.

land. And the same blockade kept out of the South the manufactured goods which had always been imported.

After the war, people started farming again. But many southern leaders thought that the South would make better use of its resources and be better off if business and industry were developed along with agriculture. As soon as there was capital wealth for investment, factories were built for processing cotton, lumber, and tobacco. Railroads were built to haul the finished products to markets. After 1880, manufacturing in North Carolina grew rapidly.

Do You Know

1. Why the first cotton mills were located in the Piedmont section?

2. Whether the processes in the manufacturing of cotton goods by machines are in general the same as those used in manufacturing cloth by hand—carding, spinning, weaving, etc.?

3. Why the early cotton mills were local industries?

4. What was meant by saying that cotton was king in the fields but not in the factories of the South?

5. What the first industries in your community were, and how they got started?

Have You Read

What your history books tell about the beginning of manufacturing in North Carolina? See references at the end of Chapter 28 and also:

Arnett, *The Story of North Carolina*, pp. 376–380.

Newsome and Lefler, *The Growth of North Carolina*, pp. 339–352.

Warren, *North Carolina Yesterday and Today*, pp. 229–234, 273–280.

30. *Resources for Industry*

WHAT makes it possible for manufacturing to develop, in North Carolina or anywhere else?

To begin with, there must always be capital wealth to pay for buildings, to buy machines and materials, and to pay wages and salaries. But a factory cannot be located in a place just because someone with capital to invest likes the country or the climate, or thinks a factory would make his home town more prosperous. There are certain other essentials to manufacturing which men must look for when they are thinking of building a factory.

North Carolina has many of these other things necessary to manufacturing. Besides capital, North Carolina has raw materials, power, railroads, highways, and banks. Some of these resources are natural, and some are man-made. All are resources from which the state creates wealth.

How is North Carolina using these resources? How does the manufacturing we have compare with the manufacturing we could have?

The answers to these questions may point the way toward greater wealth for the state. For North Carolina cannot be satisfied with knowing that it has resources for manufacturing and that these resources are being used. The state must make sure that its resources are producing as much wealth as possible.

We must measure our industrial resources just as we did our agricultural resources. We must measure the wealth produced by industries using these resources. Then we will know whether we are making the most of what we have.

Among our resources for industry are people. Machines today are often operated automatically. They do mechanically some of the jobs which used to be done by men and women. But our manufacturing still needs people. There must be people to start machines and regulate them and stop them. There must be people with hands to lift and carry, to push levers and turn valves. There must be people to watch machines at work, to listen to the hum and clatter, and to know each sound so well that any change is like a voice saying that something is wrong.

Our human wealth is one of the reasons why manufacturing has developed in North Carolina. People—thousands and thousands of people—are one of our great resources. We have workers for mills and factories, people to do things and make things.

In the summer of 1946 there were almost 360,000 wage earners working in the factories of North Carolina, almost 360,000 pairs of hands.

What do factory jobs mean to all these people? In 1946 these jobs meant for the workers an average income of about $1,765 a year, if they worked full time throughout the year.

Some of the people in North Carolina factories do unskilled work. Many others do work that is only partly skilled. The pay for unskilled or partly skilled work is never high so long as there are a lot of people wanting jobs. And North Carolina has plenty of people to work in factories. There are probably enough to double the number of wage earners now in industry.

Who are all the people ready to work in North Carolina factories?

There are the children of factory workers. When they grow up, they often take jobs alongside their mothers and fathers. Some of the men and women who have been farming all their lives are ready to try something else. Young people may be eager to leave the farms where they grew up if they have seen their parents working hard without much to show for it.

All these people often work at moderate wages. Even when the pay is low, it is probably more cash money than they are used to having.

How can North Carolina's workers in industry earn more money? Some of them have joined groups which are trying to help their members make a better living.

One of the state's great resources for manufacturing is its people. Factories need intelligent workers with skillful hands.

These groups, which are called unions, have been formed in many parts of the country where there is a great deal of manufacturing. Workers hope their unions will arrange with factory owners to raise the pay of labor and improve working conditions.

But many workers in North Carolina are unskilled or only partly skilled. Where can they learn the skills which might make it possible for them to earn higher wages? And even if they learn these skills, where will they find jobs?

These are not easy questions to answer. The way things are now, most beginners in North Carolina industries have to learn by doing. They are trained in the factory where they hope to get jobs. They start out as helpers, learning from experienced workers, and they are paid low wages while they are learning. And since a factory cannot afford to teach anyone how to do work for which it has no use, beginners learn only the particular skills required to do the jobs for which the factory needs workers.

Men and women in industry must be trained to do things and to make things. They must know how to work with machines.

Most of the jobs in North Carolina industries require only limited skills, and workers learn only enough to do these simple jobs. They seldom have a chance for anything else.

So here in North Carolina we have plenty of people, people who are among the important resources for manufacturing. And what about our other resources? Do we have the other resources that we need for industry?

Have You Learned

1. What is necessary to carry on manufacturing?

2. How many wage earners there are in the factories of North Carolina?

3. What the average income of a worker is?

4. Why North Carolina does not have more skilled workers?

5. How North Carolina can be rich in human wealth while many of the people who make up the human wealth are poor?

Have You Read

Kelly and Baskerville, *The Story of Conservation in North Carolina*, Chapter 22?

31. *Delivering the Goods*

TRANSPORTATION is another "must" in the development of industry. There has to be some way of hauling raw materials to factories and of taking manufactured goods to market, some way to deliver goods.

Factories used to be built near the sources of the raw materials they processed. North Carolina's three leading industries started in this way—because the state grew cotton and tobacco and had large areas where trees could be cut for lumber and for the manufacture of wood products. These early factories processed local materials and sold many of their products to local markets.

Then men built more roads and more railroads. And as transportation improved, raw materials could be hauled from a distance to factories built where other conditions were good. Today factories in North Carolina still use cotton and tobacco and lumber produced in the state. But trucks and trains and ships bring cotton from other southern states and from Egypt; tobacco from Greece, Bulgaria, Palestine, Syria, Turkey; lumber from Tennessee, Arkansas, Mississippi, and Alabama, and from Mexico, South America, and the Philippines.

Factories processing a raw material that is bulky and heavy and hard to handle—iron ore, for example—are

Railway freight cars are stopped alongside the loading platform of one of North Carolina's large mills, and boxes full of towels are wheeled out of the doorways into the cars. Another state product is on its way to distant markets.

generally built near the source of this material. Hauling it any distance is often difficult and expensive. For other industries, the nearness of raw materials is not so important today as it was in the past, because we have ways to haul the materials cheaply.

Factories also need good transportation for delivering their products to markets. People used to say that if a man made a better mousetrap than anyone else, the world would beat a path to his door. They meant that the maker of a good product does not have to worry about selling it; it sells itself. But advertising today is very persuasive. And there are a great many men making "mousetraps" and telling everyone what good traps they are—whether they are or not. So a manufacturer cannot always wait for the world to come to him. Selling is as important as manufacturing.

In order to sell what he makes, a manufacturer must be able to deliver it—easily, quickly, and cheaply—to the place where the buyer wants it, at the time when the buyer wants it. That is why a manufacturer needs good transportation facilities. He must have highways and railroads over which he can reach markets everywhere, ships to carry his products to all parts of the world, airplanes to carry products which are perishable or very light in weight.

With modern transportation, factories are no longer limited in their markets to neighborhood buyers or people living at a distance of only a hundred miles or so. North Carolina industry has unlimited markets. The large population centers of the east—in Washington, Philadelphia, and New York and up in the New England states—are not far away. And beyond them are the markets of the world.

Today North Carolina has two ports which ships of limited size can enter. There are about 400 miles of inland waterways along the coast from Currituck Sound to

North Carolina has more than sixty thousand miles of roads over which the people who sell things can reach the people who buy things.

Boxes of flowers are carried from a truck into a cargo plane waiting in the hangar of an airport. Daffodils picked in North Carolina one day will be on sale in Chicago the day after.

Waccamaw River. There are more than 5,000 miles of railroads in the state.

North Carolina has about 11,000 miles of main highways and nearly 50,000 miles of other roads. It cost the state millions of dollars to build the highway system, and every year the state spends large amounts of money repairing the roads and building new ones. Funds from the federal government are used on the federal highways in the state.

There are a number of airports in various parts of North Carolina. Some of these are used by air lines as regular stopping points for taking on and letting off passengers. There are also freight planes and air express for carrying North Carolina products to market.

So here in North Carolina we have people to make things. And we have roads and railroads, waterways and airways. We have ways of delivering the goods that people make.

And what about power? Do we have the power we need for manufacturing?

Making These Maps Will Be Helpful

1. A large map of your county showing highways, railroads, airlines, navigable rivers, etc.

2. A large map showing these facts about the state.

3. A map of the United States showing distances from some manufacturing city in North Carolina to key population centers in the United States.

4. A world map showing where the chief raw materials imported by North Carolina come from, and where products manufactured in North Carolina are sent.

The Maps Will Help You Answer These Questions

1. What part of your county does not seem to have adequate transportation services? What things should you take into account when deciding whether an area has all the transportation services it needs?

2. What have you already learned which helps to explain why certain parts of the state have less in the way of transportation than others? While you are thinking about this, look at the map in the front of the book showing the physical features of North Carolina.

3. What method and what route would it be best to use in sending manufactured goods from your community to New York? To Boston? To Cincinnati? To London?

Other Books Will Increase Your Understanding

Arnett, *The Story of North Carolina*, pp. 254–263, 277–282, 380–382.

North Carolina: A Guide to the Old North State, pp. 64–70.

Newsome and Lefler, *The Growth of North Carolina*, pp. 193–200, 214–215, 376–387.

Rowe, *Discovering North Carolina*, Chapter 12.

Warren, *North Carolina Yesterday and Today*, pp. 157–166, 285–290.

32. *Power and Industry*

ANOTHER essential of manufacturing is power—some way of turning the huge wheels of industry.

Our great local source of power is in our rivers. As the water in the rivers falls from the high mountains toward the lower lands lying to the west and the east, it is converted into a giant of power—electricity. The conversion of water power into electric power is in itself an important industry in North Carolina. Its product, power, can make it possible for other industries to develop in the state.

A man who did early manufacturing, when things were really made by hand, had only the power of his own muscles to work with. The things he was able to do were fixed by the limitations of his own strength. He might dream of tools to do the work of many men, but until there was a way of getting more power than the energy of one man, such tools would never become realities.

When heavier tools and machines were invented, they were operated by the combined strength of several men working together. Increasing the number of workers was the simplest way of increasing power. Then men began using the strength of domesticated animals larger and stronger than themselves—oxen, camels, horses, donkeys, or mules.

Men built windmills and spread sails of cloth over the long arms of the wheels to catch the power of the winds. And they built mills where water moving with the tide or water falling from a higher to a lower level turned wheels and generated power to operate machinery.

In these ways the ability of a man to lift things, make things, and do things was enlarged beyond the limits of his own strength. For centuries these were the chief sources of power men had to help them with their work. Then they began to experiment, trying to find something stronger—power produced by burning fuel to make steam.

Early in the eighteenth century an engine was built which burned coal to generate steam for power. Before the end of that century a Scotsman named James Watt had improved the engine so much that it used only a fourth as much fuel as the old one to do the same amount of work. This engine could operate many kinds of machines.

After the work of James Watt was finished, other men continued to experiment with the steam engine, making it better in many ways. Men had found a powerful helper.

Machines can never do work beyond the limitations of available power. Machines operated by only men or animals cannot do very heavy work because the power is small. The power of wind is stronger, but it fails completely whenever the winds stop blowing. Water power is greater, but it limits the location of factories to the waterside. And water power is not extremely efficient; it cannot be used for very heavy machinery.

With steam, however, available power became so great that men had to change all their ideas of what machines could do. It was the beginning of great industrial change—a mechanical revolution that made the world a very different place in which to work and live.

All the various sources of power have been used in North Carolina in the past, and are still being used in the state today. The first settlers had little more than the strength of their own hands to work with. Later, men used animals and wind and water. When they started to develop cotton mills into a regular industry, they located their factories in the Piedmont by the side of rivers falling from the mountains to the lower levels of the plain. For a number of years most North Carolina industries depended on this direct water power.

After 1880, many of the cotton mills used steam power instead of water power. Coal to produce the steam had to be brought from other states, and this was expensive. But in the end the final cost of production was cheaper. With water power, mills sometimes cannot operate even if there is work to be done, because whenever the rivers are low, there is no water to turn the wheels. Steam made it possible for machines to be operated steadily, and mill managers could plan their work efficiently on regular schedules.

Since the state has no large coal deposits, many factories that might have been built in North Carolina were built somewhere else nearer the coal that was used for power. North Carolina's industrial expansion had to wait for the development of another kind of power, electricity.

Lacking large coal deposits suitable for industrial use, North Carolina nevertheless has another great natural source of power. It has water falling from one level to another.

Early in the nineteen-hundreds a new way of converting this energy of motion into power was developed in North Carolina. The water was "harnessed" and made to turn wheels, as it had before. But power from the

Early mills were built at the water's edge. They used the direct power of falling water to turn their wheels and run their machines.

Today it is the power station for generating electricity which is built at the water's edge. Wires carry the power through the country to factories and mills.

turning wheels was no longer used to operate machines directly. Instead, the energy was transformed into electricity. And this electric power, carried by wires to factories sometimes miles away, was then used for the operation of machines. Throughout the 1920's, industrial development in North Carolina followed the growth of electric power.

At first this power was hydro-electric power—called this because it was produced by the water in the state's rivers. *Hydro* is from the Greek word meaning *water*. To use these river resources, companies dammed rivers at the fall line and above it, and built plants to transform the flow of water into electricity.

In 1933 the federal government established the Tennessee Valley Authority, which built dams on certain rivers—some of them in North Carolina, west of the mountains—and converted water power into electricity.

Many factories began using electricity if they were near enough the generating plants for this to be practical. If they were far away, the electricity which "leaked" out of the transmission lines made hydro-electric power too expensive.

But factories far from hydro-electric plants wanted to be able to use electricity. They were at a disadvantage if they couldn't. And people in towns and cities everywhere wanted the conveniences of electricity in their living and working. So scientists set about trying to find a way to produce electric power economically in places where there was no flow of water.

Before long these men had found other ways to produce electricity. One system burns coal to produce steam, and uses the power in steam to generate electricity. Another system produces power by burning oil in Diesel engines.

In some parts of the country, gas is used to produce electric power.

North Carolina has electricity produced by water and it has electricity produced by fuel. In the extreme western part of the state most of the power is hydro-electric. In the Piedmont much electric power comes from water, but much of it also comes from steam. East of the fall line there are no hydro-electric plants; most of the electric plants there are steam. Scattered here and there in each section are a few plants which burn oil to make electricity.

In 1945 more than half of the electricity in public use in North Carolina was generated by coal or oil. The rest came from water power.

Many factories in North Carolina have their own plants for generating the electric power with which they run their machines. If these generating plants ever get out of order, there is a substitute ready—power from public utility companies. This supplementary power means that in an emergency, work does not have to stop just because a factory's own power plant is broken. The machines can keep on without interruption.

Industry had already seen one mechanical revolution. Steam freed factories from the side of rivers and gave men power strong enough to operate gigantic machines wherever they wanted to.

Electric power brought about another mechanical revolution as great as the other one had been, perhaps greater. Electricity freed machines much as steam had freed factories. Electric power did away with the revolving shaft and the belts necessary with direct steam. Electric power made it possible for motors to be built into machines and machine tools, as part of them. The result is the modern assembly line with its streamlined production.

Be Sure to Find Out

1. What kind of power is used to operate the machine shops and factories in your community now.

2. Whether these shops and factories ever used a different form of power. If so, why did they change?

3. Whether there are any stations for generating electricity near you. Whether these stations use water or fuel to generate power.

4. What company or group furnishes electricity to your community.

5. What is meant by the statement on page 247, "Electricity freed machines much as steam had freed factories."

6. What have been the "highlights" and the "highlighters" in the development of power in North Carolina.

7. What connection there is between the development of hydro-electric power and the conservation problems and practices you learned about earlier.

Be Sure to Read

Kelly and Baskerville, *The Story of Conservation in North Carolina*, Chapters 8 and 9.
Newsome and Lefler, *The Growth of North Carolina*, pp. 388–389.
Rowe, *Discovering North Carolina*, Chapter 9.
Warren, *North Carolina Yesterday and Today*, pp. 273–276.

33. *Textiles*

THE small cotton mills built along the streams of the Piedmont during the early years of the nineteenth century were forerunners of an important part of North Carolina today—the textile industry.

It is called the textile industry now because the mills use other fibers besides cotton. They use other natural fibers—silk or wool. And they use artificial, man-made fibers—rayon or nylon. The word *textile* is a general term meaning a *fabric that is woven*, without saying whether it is woven out of cotton, silk, wool, rayon, nylon, or some other fiber.

Up to 1850 there were at least seven cotton textile mills in North Carolina. In 1939 there were 695 active textile mills. In 1946, according to one authority, 255 new textile industries were established in North Carolina. That year more than 217,000 wage earners worked in the textile mills of the state—60 per cent of the wage earners in all our industries. That is how important the textile mills have become to the people of North Carolina.

North Carolina has large textile centers: Gastonia, Charlotte, Winston-Salem, Kannapolis, Durham, Greensboro, and others. But even today we have no large industrial cities which compare in size to those in many

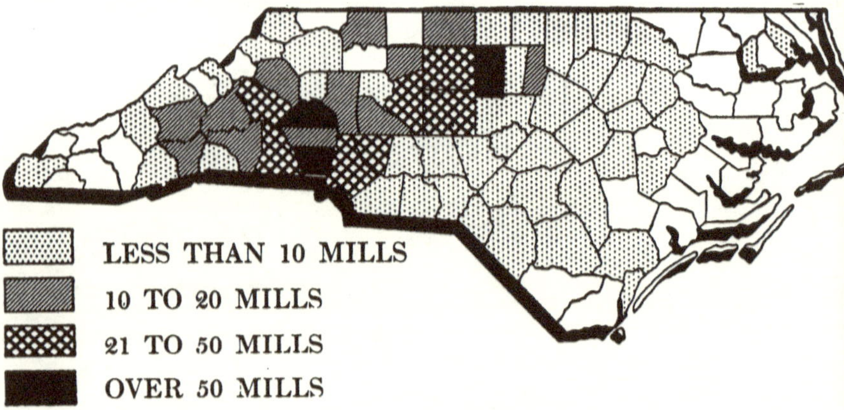

TEXTILE MILLS IN NORTH CAROLINA

other industrial states. Most of our mills are scattered throughout the Piedmont, in towns which are sometimes within sight of each other.

These industrial centers have many of the advantages of large cities. They have banks, railroads, and highways. They are near sources of commercial electric power. They are near the fields which produce some of their raw materials.

And they have an additional advantage. There are plenty of workers. And these workers can live on farms or in small communities near the mills, instead of crowding into closely-packed cities.

This is an advantage to the mills. And it is an advantage for the worker, too. He has a chance to live a suburban or country life and still be close to the factory where he works. He can combine many of the benefits of city and country.

A list of some of the things made by our textile mills shows how the industry has grown since water in North Carolina rivers started turning spindles:

Cotton yarn and thread: combed, carded, mercerized, spun

Cotton fabrics and woven goods: unfinished "gray goods" which still has to be bleached and dyed; sheeting; toweling; suiting (especially denim); colored cloth; canvas for awnings, tents, sails, and covers

Knit goods: full-fashioned hosiery; seamless hosiery; underwear; bathing suits; sweaters; tubing; cloths

Household goods: sheets; pillowcases; bedspreads; towels; blankets; rugs; wash cloths; curtains; draperies; upholstery fabrics

Braids, cordage, twine: sash cords; ropes; string

Cotton waste and cotton batting

Bags: burlap and jute; cotton bags

Miscellaneous: surgical gauze; brake linings; elastic webbing; corduroy

These are some of the things made on North Carolina spindles, looms, and other textile machines.

Sometimes we brag a little about North Carolina. We tell people that we have the largest underwear mills in the United States, the largest towel mills in the world,

Cotton from the opened bales is loaded onto breaking machines (in the background of this picture). The workers are careful to take cotton from several different piles at one time. This mixes the fibers and makes a more even thread.

Complicated machines like this do the carding which a woman used to do by hand when she was getting ready to spin in her home.

the largest denim mills in the world, and the largest hosiery mill in the world. We often say that we have more active spindles making yarn than any other state.

There are also factories doing work closely related to the textile industry. A few mills do bleaching and dyeing. Others make clothing and various finished articles out of fabrics woven on North Carolina looms. Some of these finished products are: men's and boys' work shirts, dress shirts, shorts, sportswear, neckties, trousers, and overalls; women's and girls' dresses and underwear; handkerchiefs; belts; shoe laces; gloves.

Fabrics in some of these articles contain silk, wool, or rayon, but cotton is the real king of the textile industry in the state. And although North Carolina now has a long list of textile products, yarn is still one of the most important—just as it was when the early mills opened for business.

A cotton mill's size is measured today not only by how much yarn the mill makes but also by how many spindles it has in use. A spindle holds the bobbin or quill on which the yarn is wound as it is spun. In 1939 North Carolina mills had 5,711,000 active spindles—one fourth of all the spindles in the United States, and more than any other state. This means that North Carolina does a great deal of the most elementary part of textile manufacturing.

The textile industry uses many of the state's resources. It gives work indirectly to thousands of North Carolina farmers by buying and processing some of the cotton they grow. In 1946 it employed many salaried workers and more than 217,000 wage earners.

What does the industry mean to these thousands of people? Does it make the best possible use of raw materials and other resources? How much wealth does it produce?

In 1939 the total output of the textile industry in North Carolina was worth about $550,000,000. The most valuable products were

cotton broad woven goods	$217,000,000
cotton yarn	100,000,000

After cotton has been carded, it is drawn together by machines into soft fluffy ropes. These ropes go through a process of twisting and drawing and straightening, all the time becoming stronger and finer.

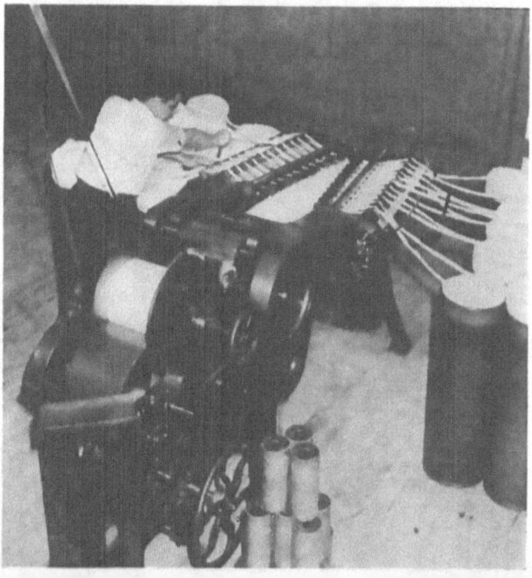

full-fashioned hosiery	$58,000,000
rayon broad woven goods	53,000,000
seamless hosiery	52,000,000

The value of a manufactured product is only one way of measuring the value of an industry. There are other yardsticks by which it may be measured. One is based on the number of wage earners and the amount of wages, another on the number of salaried persons and the amount of salaries, still another on how much value is added in manufacturing.

Some of these yardsticks are a little difficult to read and understand. Their meaning may be entirely clear only to older, experienced students of economics. But there is one yardstick which is comparatively simple. How much money is a factory worker paid during a year? What is the average wage?

Of course some wages are higher and some are lower than the average. There may not be a single worker who is actually paid the average. Nevertheless the average

This machine is finishing the job of spinning. The worker is doffing—that is, he is removing bobbins from the bottom row as soon as they are full of thread. These bobbins are ready to be used in shuttles on the looms.

Threads used for the warp of cloth must be taken off the bobbins. In a slashing machine the threads are passed through a mixture which stiffens them. Then they are wound on wide rollers, called beams.

wage per wage earner tells something about the value of an industry in a language everyone knows. It is a convenient yardstick for measuring an industry.

In 1939 the 695 textile mills in North Carolina paid approximately 181,000 wage earners about $133,000,000. This is an average annual wage of $734.

In the whole South in 1939 the average pay of the worker spinning cotton yarn and thread was only $610. The average pay of the worker making cotton woven goods was $729. In North Carolina the average was not very different.

Low wages in many textile jobs are paid in many other states besides North Carolina. They are not limited to our state alone. As a rule, spinners and weavers are not highly paid anywhere in the United States. In 1946 the textile workers in North Carolina earned an average of $33.67 a week. In the United States the average was $36.99. There is very little difference between the two figures.

Why are wages for spinning and weaving lower than

This is a very complicated loom, which has a double warp instead of the usual single one. The shuttle, which carries the thread back and forth through the warp, cannot be seen in this picture.

in many other jobs? The reason is this. The tools invented by men to help with their jobs are now doing the important part of the work automatically. Instead of men working with the help of machines, many machines in the textile industry work with the help of men. Men and women without much skill or training are able to operate the spindles and looms of today.

The reason may be easy to find, but the cure is not. How can these workers earn a better living? No one would destroy the machines which have made work easier and living pleasanter for all of us. But how can textile mill owners pay higher wages and still sell their products at prices which people are willing and able to pay?

Perhaps, in order for the cotton mill worker to make a better living, North Carolina needs mills producing finer, better quality, higher priced goods. Perhaps North Carolina needs more mills where the machines, instead of doing the work almost alone, must be operated by skillful men and women, mills where a worker is actually worth more.

Our textile mills have already begun to expand into a more balanced industry. They no longer make only coarse materials or do only the elementary steps in manufacture and then send semi-finished products somewhere else to be finished. The mills are beginning to make finer yarns and materials. They carry on the more difficult processes of manufacture, and turn out products ready for consumers.

This kind of expansion in the textile industry means a better and more profitable use of North Carolina's resources. It means higher wages for North Carolina workers when they are no longer mere machine tenders, when they are doing difficult jobs requiring knowledge and training and skill.

Our textile industry has by now outgrown the cotton crop which was one of the reasons why mills were first built in the state. In 1940–1941, North Carolina mills processed about 2,832,000 bales of cotton, while our farms produced only about 693,000 bales. The industry used about four times as much cotton as the farmers grew.

With the mills needing so much cotton, we would expect to find that they bought all that was grown in the state. But some North Carolina cotton has been a low-grade, short staple which our mills could not use. They have been buying only about half the cotton crop grown in the state. The rest of the American cotton they use has been bought from other southern and southwestern states, especially from the states of the Mississippi Valley. In some years over 40 per cent of the American cotton used by North Carolina mills came from the rich delta lands of the Mississippi Valley where farmers grew long-staple cotton.

Farmers in North Carolina have recently improved the quality of their cotton. They are planting seeds of tested quality and growing less of the short staple, more of the longer. Since a better grade cotton brings better prices, the farm income is increased. And mills in the state can buy more of the cotton grown at home. This means a better balance between the cotton field and the cotton mill.

Do You Know

1. Where most of the textile mills are located? Why they are located where they are?

2. Whether any of the things listed on pages 250–251 are made in your community?

3. What the yardsticks for measuring industry are and how the North Carolina textile industry ranks according to these?

4. What must be done if cotton mill workers are to make a better living?

5. Why North Carolina imports so much of the cotton used by her cotton mills?

Why Don't You

1. Collect samples of the textile products manufactured in the state?

2. Go exploring to see how many articles made in North Carolina are on sale in your community?

3. Make some designs suitable for dress or other fabrics?

4. Read Chapter 10 in Rowe, *Discovering North Carolina?*

34. *Tobacco*

TOBACCO has been raised as a cash crop in North Carolina ever since the land above Albemarle Sound was settled. Colonists grew tobacco, dried the leaves and packed them in hogsheads, and shipped them out of the colony to market.

The grower, if he chewed or smoked or used snuff, kept enough of his crop for his own needs. At first he did nothing to the tobacco except twist a few leaves together into a "rope" from which he broke off whatever he wanted. Then he began adding various flavors to suit his taste, such as spice and oil, licorice, sugar, and rum. In order to press this mixture into shape, he bored holes in a log and plugged the holes with tobacco. Later he split the log and took out the "plug" ready for use.

When a man had more than he needed, he sold or traded his surplus plugs to neighbors who did not raise tobacco. A tobacco farmer often realized that he was making a good income from selling his mixture. So he bought tobacco from other farmers, and spent more of his own time manufacturing than growing.

Factories sprang up in almost every little village in the North Carolina Piedmont counties bordering on Virginia. They were not factories as we know them today. Many of them were old, made-over barns. The manufacturing processes were not complicated. Shaping of the tobacco plugs was done with hand presses instead of by packing it in the holes of logs, but the equipment was simple. In addition to presses, there were only rough tables and work benches and large kettles.

By 1850, according to a census of manufacturing, North Carolina had 81 tobacco factories employing 621 workers, with products worth $308,630. As this census did not include any business worth less than $500 a year, many small tobacco factories were probably not listed.

In 1860 the value of tobacco products was almost four times greater than in 1850, and the number of workers had more than doubled. There were still less than one hundred factories. One made cigars. Some made both smoking and chewing tobacco. But most of them made "plug" and nothing else.

In 1865 Washington Duke, whose name was to become great in North Carolina's tobacco industry, began the manufacture of smoking tobacco. In a small barn he pounded tobacco leaves with sticks, sifted the tobacco, and put it into small cotton bags. The bags, with some plug tobacco, were loaded into a covered wagon. Food and other necessities were packed up. And off went Washington Duke, camping out at night as he peddled his products through neighboring counties to the east.

Other small factories appeared in North Carolina after the War between the States, and their tobacco wagons by the hundred jounced along the rough country roads. Duke's story in those years is typical of North Carolina's early tobacco manufacturers. It was what the Dukes did with their business later on that made them different from many of the other manufacturers.

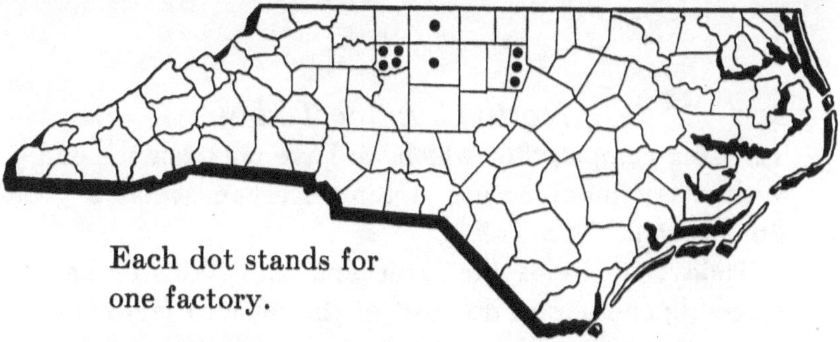

TOBACCO FACTORIES IN NORTH CAROLINA

By 1880 North Carolina had 118 tobacco factories with over three thousand employees. The products, valued at over $2,000,000, were chewing tobacco, pipe tobacco, some cigars (mostly "little" cigars), and snuff.

Soon after 1880, two Durham companies began to manufacture a product which in the years to come would turn North Carolina into the tobacco capital of the world. They started making cigarettes.

Cigarettes had been smoked in England and on the continent for some time, but they were considered effeminate in this country until after the War between the States. Then they became popular over here.

The first cigarettes were made by hand. About three hundred Poles were brought to Durham—skilled workers so expert each one could roll as many as 2,500 cigarettes a day, if he worked ten or twelve hours.

Then James Bonsack, a Virginian, invented a machine to do the job which had been done by hand. His machine could make as many as 100,000 cigarettes a day. In 1884 these machines were installed in both the Durham factories which had been making cigarettes by hand, and before very long other companies in the state started turning out the new product.

With cigarettes tumbling from machines in quantities never before dreamed of, the chief problem was to sell as many as could be made. And thus began a great advertising campaign of which we have not seen the end—with every manufacturer telling everyone what a good "mousetrap" he makes.

Today, in Reidsville, Durham, and Winston-Salem, three big companies do most of the tobacco manufacturing in the state. Chewing tobacco and smoking tobacco, snuff, and a few cigars are still made, but the chief product is the cigarette.

How does the cigarette industry rank, according to industrial yardsticks?

The cigarette industry creates a great deal of wealth. In 1939 the value of cigarettes made in the state, exclusive of taxes, was over $281,000,000. And in addition to this, the federal government collected over $250,000,000 in taxes on North Carolina cigarettes alone. In 1945 the value of North Carolina cigarettes was almost $400,000,000, and the amount of the federal tax was more than $400,000,000. The product of the state's cigarette industry helps to create wealth for the whole nation.

Cigarettes are sold in packages of twenty, and most of those made in North Carolina cost about sixteen cents or seventeen cents a pack. The federal tax of seven cents is included in this price. At this rate, factories in the state must be making a great many cigarettes if their value and federal tax amount to about $800,000,000. And this does not include cigarettes that are exported.

At the end of June 1943, North Carolina factories were turning out about 120,000,000,000 cigarettes a year—about half of all the cigarettes manufactured in the whole country. This means that if the factories worked every day except Sundays, they made over 383,000,000 cigarettes a day. And by 1946 the output had greatly increased, to perhaps 160,000,000,000 cigarettes.

Machines do most of the work in making cigarettes. Sometimes the main job of men and women in cigarette factories is to watch the machines and make sure that they do their work well.

Think how many wage earners would be employed if cigarettes were still made by hand, with each worker rolling about 2,500 a day. To make the cigarettes produced in North Carolina there would have to be more than 185,000 people doing nothing but roll cigarettes.

As it is, machines do most of the work, and very few workers are necessary. North Carolina cigarette factories have few wage earners. In 1946 there were between 11,000 and 12,000, and only a small number actually worked at the machines making cigarettes. Many of them had other kinds of jobs.

Today the cigarette machine does most of the work—the machine which started out making 100,000 cigarettes a day and which has been changed and improved until its daily production is now over half a million.

Once more the machine which men invented as a helper is doing the difficult part of the job—the part that a skilled worker used to do. Today workers help machines.

What is the average wage in the cigarette industry?

In 1946, when there were 11,700 wage earners in cigarette factories, the average weekly wages were $46.64. In that same year the average weekly pay of all industrial wage earners in North Carolina was $34.20, and the average for textile workers was $33.67 a week. According to this yardstick, workers in cigarette factories made more than textile workers, more than the average of all wage earners in the state.

Other things besides methods have changed since tobacco manufacturing began in North Carolina. In the beginning, part of a locally grown crop was processed for local markets, and for years this close connection between the product of North Carolina agriculture and the product of North Carolina industry continued. Today, however, the cigarettes made in North Carolina contain a mixture which includes tobacco from other states and countries.

Up until the Second World War, North Carolina factories were using a little over 350,000,000 pounds of tobacco a year. Some of this was burley and some was Maryland type. Around 35,000,000 pounds were imported —most of it from Turkey and Greece and a little from other Mediterranean countries. About 200,000,000 pounds were flue-cured, a tobacco which North Carolina grows and cures in quantities. But not all of the flue-cured tobacco used here was grown in the state. Quite a large part of it came from Virginia, South Carolina, Georgia, and other southern states.

Recently North Carolina farmers have been growing

between 700,000,000 and 900,000,000 pounds of this same kind of tobacco every year. Some of the North Carolina crop is used in other states, and a great deal of it is exported to other countries. Before the Second World War, more than half of our tobacco crop was exported.

Tobacco, which started out as an export crop in the early days of the colony, is still a large export crop today. The manufacturing of tobacco products, which started in North Carolina because the raw materials were close at hand, has grown from the processing of a local crop for local use into a world-wide industry.

North Carolina's cigarette industry is a source of wealth for the whole nation. The federal government has a seven-cent tax on every package of cigarettes sold—a tax included in the price of each pack and paid by every person in the United States buying cigarettes.

In order to collect this tax without difficulty, cigarette manufacturers pay the government for the tax stamps

This long row of machines, reaching almost out of sight, is packing cigarettes. When so many machines are at work in one factory, it is not surprising to know that North Carolina makes more cigarettes than any other state.

which are fixed to each package. Manufacturers get their money back when the cigarettes are sold.

In 1946 the federal tax on cigarettes made in North Carolina was over $400,000,000. Because of this, along with other federal taxes collected here, North Carolina generally ranks third or fourth among the states in the amount of all taxes paid into the United States Treasury every year. The cigarette industry of North Carolina creates part of the national wealth.

You'll Understand Better the Importance of the Tobacco Industry to North Carolina and the Nation

1. If you represent in graphs the information given in this chapter and whatever additional information you can find.
   a. A line graph could be made to show how the value of North Carolina tobacco products has increased over a period of years.
   b. Bar graphs could be made to show how the average income of workers in cigarette factories compares with that of other workers, and how the value of tobacco products compares with the value of other products manufactured in North Carolina.
   c. A circle graph could be made to show how much of the total tax bill of the United States is paid through tax stamps bought by North Carolina cigarette manufacturers.

2. If you make a poster showing the amount of taxes paid on North Carolina cigarettes each year for several years. You might make a border for the poster with covers of the different brands of cigarettes made here.

3. If you notice how much space is given in current magazines or how much time on weekly radio programs is used to advertise North Carolina tobacco products.

4. If you read pages 145–150 in Rowe, *Discovering North Carolina.*

35. *Forest Products*

THE first permanent settlers in North Carolina found the land covered with a tangle of trees to the water's edge —forests which had to be cut down before any crops could be planted.

The settlers used some of the wood for building their houses and making furniture, and of course a great deal of it they burned in fireplaces to keep warm in winter and to cook food the year round. Beyond this, the forests were a nuisance as far as the early colonists were concerned.

Men later found that trees could be a source of wealth in themselves. Tar and pitch were made from the sap of longleaf pines. They were called naval stores because they were used in building ships. Until about 1880 the production of these naval stores was an important industry in North Carolina. But men used the pines without first working out any plan for conservation or worrying about the time when the trees might all be gone. As a result, the huge forests of longleaf pines have disappeared. The naval stores industry is no longer of any importance in North Carolina.

For years, trees continued to be very much in the way. As long as good land was plentiful and cheap, most

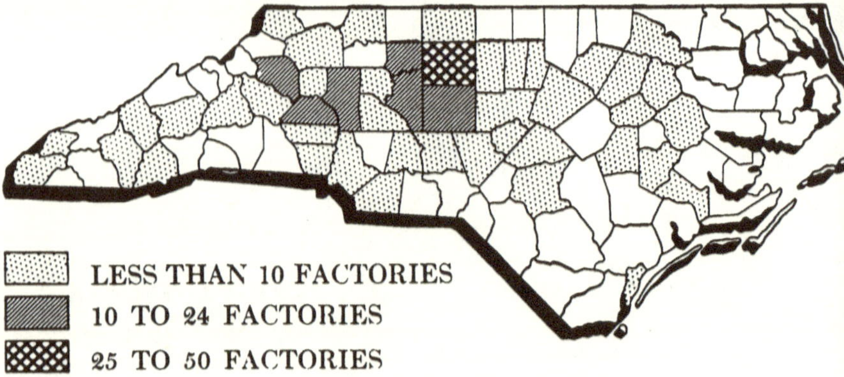

FACTORIES IN NORTH CAROLINA MAKING HOUSEHOLD AND OFFICE FURNITURE

farmers made no attempt at agricultural methods which would keep their soil fertile. They simply farmed one piece of land until it would no longer produce a good crop. Then they abandoned the old fields and cleared new land for planting. Land was cheap but men spent days and days of hard work cutting down trees.

Today there are many different uses for trees. They mean a cash income for the farmers who have wooded acres, and employment for thousands of workers in one of the many North Carolina industries where the raw material is wood.

These trees which are today a source of wealth are not different from the trees which used to be a nuisance. The difference lies in man's "know how." Increased knowledge and technological skill have found uses for a material which was a nuisance when men did not know what to do with it, or when there was little demand for it.

There are now so many forest products industries that they are often divided into three large classes: (1) Logging, Lumber, and Planing Mills; (2) Furniture and Other Finished Lumber Products; and (3) Pulp, Paper, and Paper Board.

(1) Logging, Lumber, and Planing Mills. Logging is the work of cutting down trees and getting the logs out of the forests. It is fundamental to all the other forest products industries because it supplies the essential raw material.

Some of the logging in the state is done by companies owning large tracts of forested land. Trees are harvested when they have grown to proper size, much as a farmer harvests a field crop when it is ready.

Much of the logging is done in wooded tracts on farms. The owner sells a sawmill operator the right to cut certain sizes and kinds of trees. Then a portable sawmill may be brought to the farm and the trees are sawed into lumber right on the spot. The rough lumber is either hauled to lumber yards or taken to mills where it is processed further.

In this further processing, some mills prepare wood for use in construction—rough lumber, smooth finished lumber, woodwork, and flooring. Expert craftsmen in mill workshops make all the different things that go into houses and other buildings: doors, sashes and blinds for windows, bookshelves, cabinets, cupboards, and mantelpieces.

Some mills make a business of cutting lumber into pieces of special lengths and dimensions, ready to be turned into finished products: pieces needed by furniture factories for table legs, chair legs, and other parts of furniture.

Others make all kinds of containers for shipping North Carolina fruits and vegetables to market: baskets, crates, boxes, hampers, and barrels. And because tobacco is one of the main crops in the state, several of these plants specialize in making the large flat baskets on which hands of tobacco are piled for auctioning in warehouses.

About a dozen mills in the state specialize in making shuttle blocks for the textile mills. These mills buy their wood—mostly dogwood—from local farmers, and cut it into rough blocks which are sent to small factories to be made into shuttles for the textile mills. Thus one industry exists because another industry in North Carolina has certain specialized needs.

Sometimes logs are cut into lengths and hewn into railway crossties. Or if trees are a certain size and shape, they may not be sawed up at all. They are treated with chemicals which preserve the wood, and then are used for poles and fence posts.

A few manufacturing plants produce excelsior, shingles, and plywoods.

Beautiful hardwoods are cut into very thin sheets, called veneers, which are used in furniture. Other veneers, cut from softwoods, are used in baskets for vegetables and fruits.

This part of the forest products industry—logging, lumber, and planing mills—employs many people. In 1946 there were almost 28,000 wage earners who were paid an average of $29 a week.

In the logging part of the industry, the work is heavy and rough and sometimes dangerous, but it can be done by men with little skill or training. Many of them work only part-time—perhaps a few months in winter when much logging is done. Sometimes small farmers take logging jobs in order to make a little ready money. Even though the wages are low, the extra pay helps to balance the family budget. In the sawmills there are a few skilled jobs, and in the planing mills there are many more.

North Carolina ranks among the first ten as a lumber-producing state. In 1938 our various sawmills cut 1,439,-

000,000 board feet of lumber. Of course much of this was used locally, but over a fourth of it—371,000,000 board feet—was exported to other countries or sold in other states, most of them north of Carolina and east of the Mississippi River. A little went as far away as Wisconsin and Illinois, Maine, New Hampshire, and Vermont. Most of it was sold in Pennsylvania, New York, Maryland, Virginia, and New Jersey. In these markets North Carolina's location gives the state an advantage over lumber-producing states on the Pacific or to the south.

(2) Furniture and Other Finished Lumber Products. In the early days, settlers put together their own rough benches and tables. Later, expert craftsmen set up shops where they worked at their trades, making many kinds

Cutting trees and getting them to the sawmills is heavy work and sometimes dangerous.

of beautiful furniture by hand. Between 1880 and 1890 factories in North Carolina started producing furniture by machinery. And today the manufacture of furniture is one of the important industries in the state.

Furniture factories were built in North Carolina for some of the same reasons that cotton mills and tobacco factories started here. There were people to do the work. The raw material was close at hand. Water power was available. There were roads and railroads. There was capital to be invested. And there were men with imagination who could see how to combine all these advantages.

But there was no one to remember that trees in a forest can either be harvested, like a crop, or they can be "mined" and completely used up. So trees were cut faster than others could grow. Factories were located in a place because hardwoods could be bought in the neighborhood. Now these factories have to get much of their raw material from other states, from Canada, Africa, and South America. Once more men have learned by experience that it is poor business to use a raw material recklessly without any plans for conserving and renewing it.

Today North Carolina has over two hundred factories making furniture and other finished products—most of them scattered throughout the northern Piedmont from Durham to Lenoir.

Some of these are small manufacturing plants making such things as washboards, coat hangers, dowels, spindles, and handles for mops, brooms, hoes, rakes, spades, and other tools. A number of factories make furniture and other equipment for offices, stores, and schools—such things as wooden desks, chairs, tables, counters, filing cases, and show cases. A few factories make caskets and coffins.

Over a hundred factories in the state make household furniture, wooden and upholstered. They produce tables, chairs, desks, sofas, cupboards, bookcases, beds, dressers, chests, cabinets—all the usual pieces of furniture for every room in the house.

In 1946 there were over 27,000 wage earners in North Carolina factories producing furniture and other finished lumber products. Many of these workers were making handles for tools and other miscellaneous products. The jobs require so little skill that the worker can be paid only low wages. These are low-price, low-wage industries.

In the furniture factories, workers are paid well for doing the difficult jobs that require skill and training. But for each skilled worker in this industry there are many unskilled workers who lift and carry the lumber around in the yards or on the factory floor, and help the skilled craftsmen.

In this section of a furniture factory wooden dressers are being finished and made ready for shipment. Mirror frames and other parts for the dressers hang overhead, within easy reach of the workers.

Since there are a great many more low-paid, unskilled workers than there are skilled workers, the average wage in this industry is low. In 1939, 21,376 wage earners were paid a total of $14,675,520, an average of $686. In 1946 the average weekly wage was $31.20, or about $1500 a year.

Furniture factories in North Carolina use the products of some of the other industries in the state. Veneers and plywoods are used in the construction of furniture, and excelsior and rough lumber in packing and crating it for shipment. Several plants mix paints, varnishes, stains, and lacquers—all of which may be needed in finishing furniture of different styles. Muslins and patterned fabrics woven in North Carolina mills cover upholstered sofas and chairs. A few factories in North Carolina make mirrors—from plate glass manufactured in other parts of the country—which are used on dressing tables, bureaus, and dressers.

Some furniture factories which produce their own power burn wood shavings and sawdust as part of their fuel. In this way they get rid of any waste wood and at the same time make their power cheaply. To supplement this, there is the product of another great North Carolina industry, electric power.

The furniture industry has had an important part in the industrial development of North Carolina. In 1939, the state led all others in the production of wooden household furniture. For many years much of our factory-made furniture was of a cheap grade, and some of it still is. But today more and more of the furniture made in North Carolina is of a high quality. Every year people from stores in many parts of the country come to buy furniture at the exhibits held in High Point.

Wood pulp which will be made into paper is always carefully washed with pure water.

(3) Pulp, Paper, and Paper Board. In this third group of the forest products industries the raw material is wood, just as it is in the other two, but it cannot be recognized as wood. It no longer appears in its natural form as in lumber or furniture.

The fibers which make up wood are separated into pulp, and then the wood pulp is made into many kinds of paper and paper board.

Separating wood fibers may be done mechanically—by grindstones which grind up the wood as fine as possible. The pulp produced by this method makes a coarse, low-grade paper.

Another way to make pulp is a semi-chemical method in which logs are steamed and softened before they are ground. The pulp is of better quality than pulp made mechanically.

A third method of making pulp separates the wood fibers by using chemicals. Since chemicals get rid of everything that is not pure wood fiber, the pulp made chemically is much finer and stronger than pulp made by grinding.

Because wood pulp is often produced chemically, the industry is sometimes considered a chemical industry. But the other methods of producing pulp are still used. So it is classified here as it generally is, as a part of the forest products industries.

In one of the steps between wood and paper, pulp is soft and fluffy and looks very much like cotton.

Making wood pulp and making paper used to be two parts of one industry, and this is still sometimes true. But production of pulp has become so specialized that today it is often a separate industry. Factories do nothing but make pulp. This is shipped in large thick sheets to other factories, where it is made into various kinds of paper, rayon, cellophane, film, and other plastics, or other products in which the raw material is pulp.

In 1941 North Carolina had four mills making pulp. In 1938 these mills used over 400,000 cords of wood, for which they paid about $2,500,000. In 1944, because of war needs, they used 600,000 cords.

In addition to the four pulp mills, North Carolina has about thirty plants making paper, paper board, bags, boxes, and miscellaneous paper products. Altogether these thirty-odd factories employ almost six thousand wage earners. Compared to the number of plants and of wage earners in the other two groups of the forest products industries, the number of plants and of workers in this group seems small and unimportant.

Measured by the yardstick of average wages, however,

the pulp, paper, and paper board industries are valuable to the state. In 1946, they paid average weekly wages of $50.22. This was much more than the average in the other forest products industries.

The average weekly wages for logging were $30 in 1946. In the same year the average in the furniture industry was $31.20, and the average in the pulp and paper industries was $50.22.

What is the reason for this difference in wages? The main reason is the difference in the skills of the workers. The people who make pulp and paper have to be more skilled, and skilled workers are worth more than those who are only semi-skilled or unskilled.

You'll Want To

1. Discover what equipment used in your home or school was manufactured in North Carolina out of North Carolina wood.

2. Explore your community to learn what wood-using industries it has.

3. Locate on a map of North Carolina the chief centers for the manufacture of furniture and other products made of wood.

4. Show in tables or graphs how the different classes of forest products industries compare with one another in number of workers and in average income per worker. Show how forest products industries as a whole compare in these respects with other industries.

5. Collect pictures of forest products manufactured in North Carolina.

6. Read pages 152–156 in Rowe, *Discovering North Carolina.*

7. Look through magazines for advertisements of furniture made in North Carolina. Do leading furniture manufacturers in North Carolina have trade-marks?

You'll Want to Think About

1. How the future development of forest industries in North Carolina might influence farming as a business.

2. How such development would promote conservation of natural wealth.

3. How more of the waste products from wood can be made use of—for example, odds and ends of lumber left by carpenters, sawdust, or dead trees in a woodland.

36. *Other North Carolina Industries*

THE making of *rayon* is one of the great test-tube industries which are the result of the technology of today. In this industry, wood is the raw material just as it is in the lumber or furniture or pulp industries. In the manufacture of rayon, however, pulp is chemically changed so that the finished product no longer looks like wood, feels like wood, or is used like wood. Since rayon cannot be manufactured without the use of chemicals, this is classed as a chemical rather than a forest products industry.

In the western part of North Carolina there is a mill which makes rayon yarn out of wood pulp. Special technical skills are needed to do this work, skills which the people living near the new mill had never had a chance to learn. Before the mill could be opened, skilled workers had to be brought from Holland to teach the local people what to do. North Carolina workers soon learned the necessary skills, and now there are two or three thousand of them in the mill.

All rayon used to be made from cotton linters, which are cellulose, like wood. Today about a fourth of our rayon is made from cotton, and the rest is made from wood.

Rayon started out as a material intended for people who wanted silk when they could afford only cotton. But it was not a very satisfactory substitute. It was too shiny, it stretched and lost its shape, and it wore out quickly. Rayon has been much changed and improved since those early days. There are now rayon fabrics that look like silk and wear almost as well, rayons that look like cotton, others that look like linen, and even some that look like wool. Rayon material is no longer thought of merely as a substitute for something more expensive. It is a beautiful fabric itself.

If rayon and other test-tube fibers take the place of cotton, this substitution will make a difference to workers in North Carolina textile mills, especially the people who do carding and spinning. And it will make a big difference to North Carolina cotton farmers, too, if they are not able to sell their crops because mills are using other fibers. They may start raising more and more trees as a crop, and plant less cotton.

Western North Carolina has one of the state's most important chemical industries, a factory which makes *cigarette paper* for the whole nation. Why is this a chemical industry when the manufacture of other paper is included in forest products? The paper for cigarettes is not made out of wood. It is made out of flax.

Until this factory opened, cigarette factories were using paper imported from Europe. Then, a short time before the Second World War, the North Carolina plant started production. Many people wondered what would have happened to the great American tobacco industry during the war if the factories had been dependent upon Europe for paper. But when the North Carolina plant opened, it made enough cigarette paper to supply all the cigarette factories in the nation. It also makes paper for other uses.

Fertilizer is another important chemical industry in North Carolina. A few factories in North Carolina produce materials that are used as fertilizers—fish meal and cottonseed meal, for example. But most of the fertilizer factories in the state do no actual making. They do only mixing.

Plant foods are brought into North Carolina from other states and countries and are mixed here in definite proportions. Then they are combined with something bulky—such as sand, or sand and lime—and are packed in bags ready for farmers.

Since a large part of the finished product is the sand,

Cooked peanuts are shaped into long flat cakes, wrapped in cloth covers, and fitted into the boxes of an oil press. In this picture a worker is taking cakes from the box after the oil has been pressed out.

or other filler which is heavy but neither scarce nor expensive, it would add greatly to the cost if commercial fertilizers had to be shipped any great distance. Factories are therefore built close to their markets, and sand is brought in from the neighborhood. The expense of ship ping the bagged fertilizers great distances is saved.

North Carolina buys more fertilizer than any other state in the nation; so it is not surprising that the state has many fertilizer plants.

Wages in these factories are low, because the workers are not skilled. And since fertilizer is needed mainly during the spring and summer, some workers in fertilizer factories have jobs only part of the year.

North Carolina has various *other chemical industries.* Peanuts are crushed for oil. There are cottonseed products besides meal—cottonseed oil, cake, and linters. For several years an industrial chemical company on the coast extracted bromine from sea water and produced the ethyl used in high-test gasoline. There are factories in the state making paints and varnishes; a little turpentine and resin;

drugs and medicines; toilet preparations; insecticides and disinfectants; tanning extracts; dyes; and a few other less important products.

Altogether North Carolina had about 150 chemical industries in 1946, with over 8,600 employees. Some of the workers are skilled and well-paid. Many of them, especially those in cottonseed and fertilizer plants, are not.

So although some of the chemical industries pay good wages, the average per worker for those in North Carolina is low. In 1939 it was $600 a year. In 1946 the weekly average was $33. This means that a full-time worker could earn about $1,600 a year—almost three times the average wage in 1939.

In addition to textiles, tobacco, forest products, and chemicals, there are a few *other manufacturing industries* in the state. The following list includes most of them.

Food products: bakeries; soft drinks; pickles, preserves, and other canned foods; flour, cornmeal, and other products of grain mills; sausage and meat; dairy products; ice cream; candy; potato chips, etc.

Leather and leather products: belting for industrial machines; packing and washers; leather for boots and shoes; textile mill roller coverings; horse collars, etc.

Stone, clay, and glass products: abrasives, asbestos, and ground minerals (mica, talc, limestone, feldspar); con-

Tile pipes are among the products made from some of North Carolina's mineral resources.

Shipyards along the coast use some of the state's lumber resources.

crete, gypsum, and plaster products; granite and marble; brick; tile; pottery and related products (kaolin); mirrors, etc.

Printing: newspaper printshops; commercial printing and lithographing; bookbinding, etc.

Metals and metal products: aluminum; lead smelting; machinery and machinery parts (especially textile machinery, oil-burning tobacco curers, farm implements, wood-working machines); cutlery and tool hardware; machine shops, foundries, and general repairs; sheet metal work, etc.

Transportation equipment: automobile bodies and parts; wagons; boats, etc.

These are about all the manufacturing industries in North Carolina. We have a number of different factories in the state. But we do not have many different kinds of industry.

Some Surprises May Be in Store for You

When you find how many different industries are located in your own community or near it. Divide your class into eight or ten committees. Let each committee try to find out about plants manufacturing one group of products mentioned in this chapter. Let each member observe as he walks or rides about for a week and let him watch local newspaper advertisements. One member from each committee might be sent to interview the secretary of the nearest chamber of commerce.

Some New Ideas May Come to You

1. When you think about and discuss in class these questions:

   a. How do the number and the size of these miscellaneous industries compare with the number and sizs of textile mills, tobacco factories, or forest industriee in your community? How do you account for the difference?

   b. In what way is the rayon industry connected with forest products industries? With textile manufacturing?

   c. Why is the yearly average income of chemical workers so low?

   d. What might be the possible effects of an expanded rayon industry on cotton farming?

2. When you read and report on such interesting books as:

   a. Huff, *Twenty Careers of Tomorrow*

   b. Leyson, *Plastics in the World Tomorrow*

   c. Mersereau, *Materials of Industry*

37. *What Factory Jobs Mean to Workers*

IN 1939, when there were over 3,200 industrial plants in the state, the number of *different* industries in North Carolina was small. According to the Census of Manufactures, there were 446 different kinds of industry in the whole United States. Of the 446, North Carolina had only 191.

This is a small number compared to some other industrial states which have about the same number of factory workers as North Carolina. Indiana, for example, had 368 out of a possible 446. California had 377. Even Virginia, with only about half as many factory workers as North Carolina, had more kinds of industry: 209.

This means, of course, that factory jobs for the workers in North Carolina are limited to the 191 industries in the state. Since some of these 191 industries are themselves very small, with few factories and few workers, there is naturally a concentration of workers in the larger industries of the state.

The table on page 287 lists the chief industrial groups in North Carolina and shows the number of wage earners and the average weekly wages in August, 1946. At the head of the list are textiles, cigarettes, and forest products. In these three groups are concentrated over 85 per cent of North Carolina's industrial wage earners.

Concentration might not be undesirable if it were in industries paying high wages. But this is not the case. The average yearly wage of North Carolina industrial workers is low. It was only $737 in 1939, when the average for the whole country was $1,038.

North Carolina wages are not high compared to other states in the South. In number of wage earners North Carolina heads the list of the eleven southeastern states. But in average wage per wage earner North Carolina is the "middle" state, sixth from the top and sixth from the bottom.

Even when we know the average wage in North Carolina, we are a long way from knowing what this means to the workers. What do factory jobs mean to the people who have them? There are probably a great many answers to this question, as many answers as there are individual workers. The real value of the money a man or a woman earns depends on how it is used, and this varies with each person.

In some factory jobs, machines seem to do more work than people.

What may be enough for one isn't enough for six or seven more. How many others must be fed and clothed and housed with one worker's wages? What kind of food does the family eat? Just enough to keep from feeling hungry, or the kind that builds sound bodies and bright minds? How many warm clothes do they have for the winter? Are there shoes for all the children? Are there medicines and doctors when someone is sick? Do they have any amusement, any fun? And so on.

WHAT FACTORY JOBS MEANT TO NORTH CAROLINA WORKERS IN 1946

INDUSTRY	NUMBER OF WORKERS	AVERAGE WEEKLY WAGES
Textiles	217,300	$33.67
Tobacco	15,400	39.47
Logging, etc.	33,300	30.00
Furniture, etc.	27,200	31.20
Pulp, etc.	5,800	50.22
Stone, etc.	4,200	32.18
Food	17,300	30.00
Clothing	7,700	26.27
Metals	3,000	36.69
Machinery	4,200	40.63
Transportation Equipment	2,800	51.74
Miscellaneous	19,200	36.04
All Industries	358,900	$34.20

There is no way here to answer these questions by adding a column and finding an average. If we had tables and charts and statistics reaching from Manteo to Murphy, we still wouldn't know what factory work has meant in the lives of all the workers in North Carolina.

However, we might be able to find out a little by asking the question not about all the workers, but just about the textile workers, since they make up two-thirds of the workers in the state. What kind of living do they make?

The average pay among textile workers is low. In 1939 it was $734. In 1946 it was about $1500. Since this is the average wage, of course many workers get more than this and many workers get less. The average pay for textile workers is less than the average cash income on the farm, which was $790 in 1939 and $2,197 in 1944.

However, a farmer's cash income generally means the work of more than one person. He and his wife both work on the farm, and sometimes their children do, too. The farm income is really a family income.

A textile worker's pay means the work of one person. His wife often works in the mill, too, just as the farmer's

The skills used in the sewing room of a mill making sheets and pillowcases are not very different from the skills needed for sewing at home.

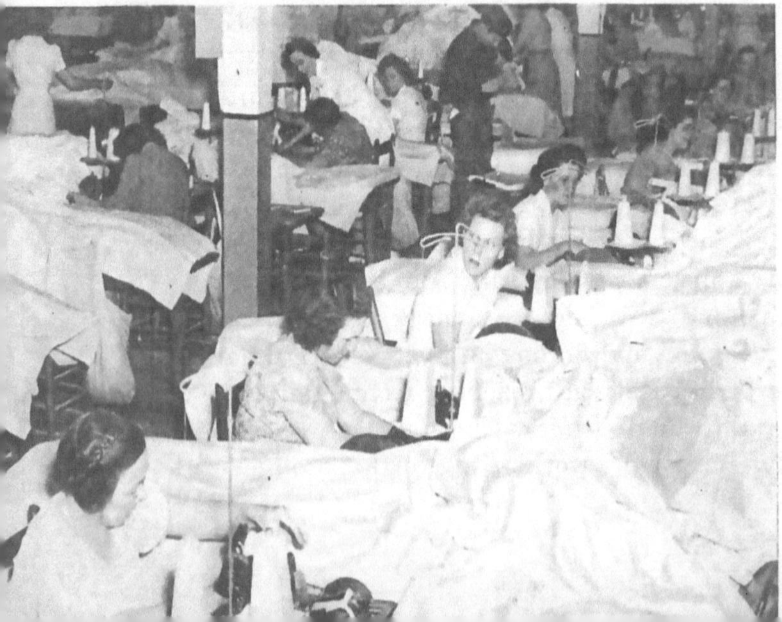

wife works on the farm, and so may their children, as soon as they are old enough. But each one of them is paid separately and individually. The family income, with two people working in the textile mills, is considerably larger than a family income with two people working on the farm.

Of course a farmer often receives "pay" which is not part of his actual cash income. If he is a tenant, he pays no cash rent for the house he lives in, as the factory worker must do. The farmer's water costs him nothing. He generally cuts the wood he needs; the factory worker generally has to buy his. Whatever food the farmer raises for his family doesn't have to be bought. The factory worker may also raise food, and often does, but he is less likely to do so than the farmer is.

These are some of the necessities for which the farmer does not pay cash. They help to make the average farm income mean a little more than it appears to, in actual money; but they do not make up the difference. The average farm income is less than the average income of a family with two or more people working in the textile mills.

Be Sure

1. To learn what is meant by the terms "concentration of workers" and "concentration of industries."

2. To show in a graph the information about the number of workers in the industries listed in the table.

3. To find what you can about the number of workers in your county employed in the different industries.

4. To discuss in class or with some adults in your community the reasons why North Carolina has fewer industries than many states.

38. *People at Work*

WALTER and Katherine Burke left the country to work in the mills. They heard there were jobs to be had, and they decided to try the factory for a while. They could always go back to the farm.

It wasn't easy at first. They got jobs all right, but they had everything to learn, and the pay to begin with was very low. They weren't used to working a certain number of hours a day under a foreman, with an inspector to check what they did to be sure it was right.

Walter and Kathie had worked hard on the farm, but they'd been their own boss. They often stopped during the day whenever they felt like it, to catch their breath, and there were winter months when Walt did practically nothing. "If I'd worked steadily on the farm the way I do in the mill," Walt says, "so many hours a day the year round, I believe I would have made pretty good money."

Some textile mill workers do piecework. They are paid on the basis of what they do. Some are paid on the basis of a certain number of work-hours during a week. If they work more than this, they are paid for the overtime.

In a way the farmer who works in his fields from sunrise to dark is paid for his "overtime," too, if his crop brings a better price because of his work. But his overtime pay depends on a big *if*—on whether he uses the time intelligently—and is much more uncertain than the textile worker's.

There are several ways in which Walt the textile worker is better off than Walt the farmer. If he is injured at his job, his bills are paid for him with accident insurance taken out by himself and his employers. When a farmer has accident insurance, he pays for it himself. Walt never thought he could afford it.

If Walt and Kathie have to stop working because the mill is producing less, they are still paid something every week for a certain number of weeks during a year. This is paid out of money the mill has deposited with the Employment Security Commission. The farmer has nothing like this.

If Walt dies before he is sixty-five years old, his family may receive a monthly income out of insurance to which Walt and the mill owner have both made payments. And when Walt and Kathie are sixty-five, they are entitled to receive a monthly income under the Social Security Act.

Walt is a weaver. His wages are a little more than the average pay in the textile industry. Kathie, who is a spinner, doesn't make as much. They live in one of the mill houses. It isn't large; but the rent is low and they have running water and electricity, which they never had in the country. They go to the moving pictures often, and in the summer time they watch the games of the mill baseball team. And they still enjoy the brightly lighted stores on Saturday nights just about as much as when they first came to town. They get a few things they need, walk around looking at everything, and stop now and then to visit with friends. With four children to buy food

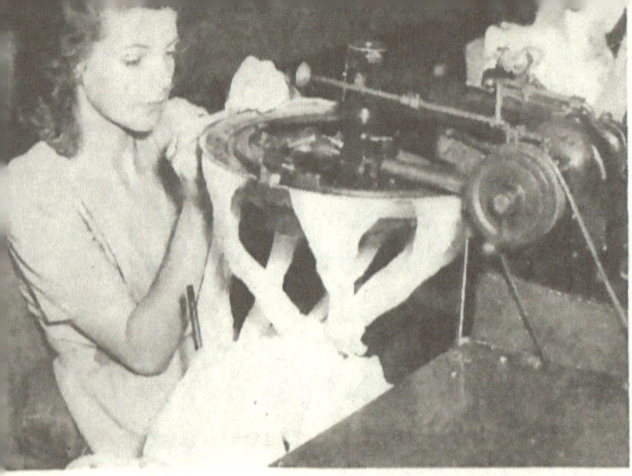

This worker is operating a machine which loops the toes of stockings.

and clothes for, as well as themselves, their money doesn't last long, and they don't have many extras.

Rose, the oldest child, has finished high school and has started to work in the mills. That will make a big difference. With her money they may be able to move to a better house. Rose is a learner; so she doesn't make much money to begin with. She'd have a hard time living on it if she weren't staying at home. But she'll make more later on.

Rose wanted to get a job in full-fashioned hosiery. "Whatever kind of work people start at," she told her mother and father, "they generally keep on doing the rest of their lives, and I want to start at something good." But there wasn't an opening in full-fashioned, or in the spinning room where her mother worked. She had a chance at a job making socks, so she took it. The extra money helps out a lot.

Henry Glass lives in another part of the mill village, where the rent is very low. His front steps are breaking away from the porch, and the roof leaks a little when the rain is heavy from the west. Henry's wages are less than the average for a textile worker.

He started out as a weaver, but he had to quit work in 1933 to have an operation, and since then he's just been doing odd jobs around the mill. His operation cost $100, and back in those days there wasn't any unemployment compensation or other help. Mr. Russell, up at the mill, paid the bill, and then took something out of Henry's pay when he got back to work—a little every week. It looked as if he'd never get it paid for. He keeps hoping he'll be weaving again soon. "A man can make a good living at weaving," he says. But the doctors tell him he isn't really able to do the work.

Henry isn't married. He lives with his mother, and his wages have to do for both of them. They generally have enough to eat, but it isn't always the right kind of food. Boiled potatoes and pinto beans and big crumbly biscuits, most of the time. Neither one of them likes meat very much, and that's lucky, on Henry's wages. As soon as Henry's mother is sixty-five years old, she'll get a little money every week—old-age assistance from federal, state, and county funds.

Andrew Dillon lives around the corner from Walt and Kathie Burke. He started working at the mill when he was twelve, and he never quit. There weren't any child labor laws in those days. He made twenty-five cents a day at first, then fifty cents. "That doesn't sound like much," he says, "but money went farther then. It would buy more. And we didn't see so many things to buy."

Andrew makes a lot more than that now. He's overseer in the yard, and weighs all the cotton that goes in and out. Two of his children have been working in the mill for quite a while; so the family has the wages of three workers.

They spend about all they earn, because Andrew and his family are living much better than they did when the children were young. They have a radio, and an electric refrigerator that cost $200. "I hated to pay out all that money," Andrew says, "but I wouldn't be without it now for anything. Or the water heater, either. It's automatic. Just turn on the faucet and the water's hot. These things make living lots easier than it used to be. Sometimes I wonder what they'll think of next."

Joe Edwards knits full-fashioned stockings. It took him several months to learn. He had some savings to live on, and after a while he was running a knitting machine himself.

Joe married a girl who was working in the mill, and she kept on with her job until they'd saved enough to start buying a place on the edge of town. "It's not very far out," Joe says. "Ten minutes after I leave the mill, I'm home. I've got a garden and some chickens, and I'm thinking about buying a cow. It all makes our money go a lot farther. And we like it out there. I figure we have the good of the town and the country, both."

Joe makes pretty good pay, but the job's worth it. Knitting full-fashioned stockings isn't easy. The machine is complicated. A man has to know what he's doing or he'll wreck it. And the work is hard on the eyes.

Walt and Kathie and Rose and Henry and Andrew and Joe—they are only a few of the industrial workers in North Carolina. They may not be "average" in every way, but in some ways they are typical of the others.

One thing, at least, they all have in common. Their chances to get jobs at good pay are limited by the fact that industry in North Carolina is concentrated, specialized. It is not that our wage rate is low. Rather it is that our chief industries are low-wage industries.

North Carolina's concentration on low-wage industries affects more than the workers themselves. It makes a difference to the whole state. When pay is low, workers have less money with which to buy what others make or sell. A low income for one group of people in the state means lower incomes for everyone. And a low income for everyone means less income for the state itself, less chance to increase the institutional wealth of North Carolina.

Since 1900, North Carolina has become the greatest industrial state in the South and one of the leaders in the nation. The expansion of industry has brought wider opportunities to many people in North Carolina. But manufacturing has been concentrated in three major industries, just as agriculture has been concentrated in two crops.

The time has come for industrial diversification if we are to use all of North Carolina's resources: the natural

Two rows of machines are knitting full-fashioned nylon stockings. Skilled workers must understand these complicated machines, and must know how to keep them running.

wealth of raw materials, mild climate, and power; the nearness to large markets; and that greatest resource of all, human wealth, hundreds of thousands of hands to do things and make things.

Do You Agree

With the descriptions of life and work given in this chapter? Why or why not?

Can You Write

A ballad or a series of poems about North Carolina workers in which you describe their work and express their difficulties and their hopes? In your literature book, *Contact*, reread Walt Whitman's "I Hear America Singing," or read some of the poems in Ruth Barnes's collection of songs of working people in America. It has the same title as Whitman's poem, *I Hear America Singing.*

Have You Seen

Any paintings by some of our modern artists picturing important industries and showing people at work? Try to secure some of the paintings by Thomas Benton, Henry Varnum Poor, Fletcher Martin, or others to display on your bulletin board while you are studying about industries.

39. *Diversified Manufacturing*

HOW can the industries already established in North Carolina be improved? And what other industries can be added to make our manufacturing more diversified? In other words, how can North Carolina's resources be used so as to create more wealth and give the people of the state a chance to make a better living?

Before answering these questions, experts first ask themselves several others. What raw materials are now being used in low-value industries when they could be used in industries creating more wealth? What raw materials are being wasted because they are shipped out of the state to be processed? What raw materials are being wasted because they are not being processed at all? The answers to these questions point toward some of the new industries which might be established in North Carolina.

And these industries which are made possible by our raw materials—what of them? Are they the kind in which people can make a better living? How desirable are they, measured by industrial yardsticks? How many workers might be employed and what is the average wage per wage earner for the same industry in other parts of the country? How many salaried persons might be employed, and how much would they be paid? How much value is added to the raw material by manufacturing? And what is the total value of the finished products?

After studying the possibilities and measuring various industries, experts have suggested ways in which North Carolina can use some of its resources to produce more wealth.

In textiles, many North Carolina mills now carry on only the first steps in processing cotton. They sell their yarn to factories in other states, where it is woven into fine materials. And they send away much of their half-finished woven goods to factories which bleach and dye and finish it. North Carolina needs plants which can do this better-quality work, plants right here in the state to take raw materials through all the stages of manufacture and send finished products to market.

A few years ago these changes might have been enough to make the most of our resources in cotton and cotton mills. Today, however, as a result of modern technology, the cotton textile industry here in North Carolina and everywhere else has a rival unknown to the mills a generation ago—synthetic fibers. If cotton is to hold its place in the world against rayon, nylon, and other synthetics, the whole textile industry must improve the quality of cotton fabrics and make them better. They must be better than they are and, if possible, better than any of the newcomers.

In this development North Carolina has a chance to take the lead. Can North Carolina technicians find a way to produce a cotton fabric better than that made anywhere else from cotton or any other fiber—perhaps one that wears longer and wrinkles less? Can North Carolina mills weave more attractive cotton materials, perhaps with better colors and more beautiful designs? If they

can, then this low-wage industry may become a high-wage industry in the state. Some progress has already been made, but a great deal remains to be done.

In the manufacture of tobacco, North Carolina is already producing for a national market. Even if more cigarette factories were built in the state, they might have difficulty selling their product. But in this industry, too, it is possible that technical experts may work out ways to improve methods of manufacturing and selling.

Many of North Carolina's forest products industries use wood in low-value, low-wage manufacturing. One improvement would be to manufacture more of the better-grade furniture. Another improvement would be an increase in the number of pulp mills. Pulp mills use the same raw material as other forest products, but they create more wealth than the others, when measured by various industrial yardsticks.

In 1944, 600,000 cords of pulpwood were used in North Carolina, but 400,000 cords were shipped away. Will the state keep on exporting a raw material which it could process with profit itself? If not, more pulp mills must be built. More pulp mills would mean greater wealth for the state and jobs with good pay for industrial workers.

With more pulp mills over the state, farmers would have a cash market for the trees from their uncleared land and the trees growing in abandoned fields and on eroded hillsides. Here is an industrial raw material which need not be exhausted, like minerals.

These are some of the ways in which experts think our leading industries might be improved.

Expansion of some of the smaller industries and the addition of a few new ones would use more of North Carolina's resources and make manufacturing in North Carolina more evenly developed and better balanced.

First, the state needs to process more foods. As things are now, North Carolina probably does not process more than 2 per cent of the canned and packaged foods used in the state. The amount of creamery butter manufactured here would provide each person in the state with only about half a pound a year. The fruits and vegetables canned would not last a single day, and prepared meats and sausage would not last a week.

Yet this is an agricultural state, already growing some foods that could be processed—certain fruits and vegetables, for instance—and able to grow more if there were processing plants to buy the crops. North Carolina farmers could follow a program of diversified farming more easily if industry in the state were diversified to include plants for canning and processing foods.

Dairy herds and meat animals have not been raised here in any quantity in the past. But the presence of processing plants, of a ready market, would make it easier for farmers to carry out one part of the agricultural program. Processing plants would help develop animal husbandry on North Carolina farms.

What do we do in North Carolina? We ship peanuts out of the state by the ton and buy them back by the pound, salted or in candy. We ship out hogs and buy them back as hams or bacon. We ship fresh snap beans to New York and buy canned snap beans from Maryland, Minnesota, Wisconsin, California, or some other state. We raise more cotton than can be sold at a profit and buy hay from the Middle West. Instead of doing this, North Carolina could be growing and processing much of this food for itself and for national markets.

Other raw materials besides pulpwood are now sent out of the state to be manufactured. Kaolin, for instance—one of our non-metallic minerals—is shipped to another part of the country. If there were plants in North Carolina for the manufacture of chinaware, the state would profit by processing kaolin as well as by mining it. Feldspar and mica are also shipped out of North Carolina. They are raw materials for manufacturing in other states.

Chemical industries may be the greatest future source of manufacturing wealth for North Carolina. The state could add many of the "test-tube" industries in which men use their scientific skill to work the magic wonders of the twentieth century.

Finding ways to use plentiful farm crops and the waste products of agriculture and industry has been one of the jobs of industrial chemists. Working at this in their laboratories, men have sometimes turned the world upside down. People today may either drink milk or wear it, changed into material that feels and looks like wool. Peanuts are bought by the sackful. They may also be bought by the yard—after the shells have been made into a heavy, wool-like fabric. Or they may be bought by the gallon—when the nuts have been made into oil for cooking. Windows of glass may be covered by curtains of glass—curtains that look like silk. Soybeans may appear on the dinner table as food, or under it—in place of leather as the soles of shoes.

Perhaps Cinderella was only a little ahead of her time when she rode to the ball in a coach made from a pumpkin.

Technological chemists have done something that sounds just as fantastic as making a silk purse from a sow's ear. They have made nylon—a thread as fine and soft as silk, and much stronger—from a mixture of coal, air, and water.

In modern synthetics, an important raw material is one of North Carolina's great natural resources: wood. Wood chemically treated becomes a fabric as soft as silk. This is rayon, which is already being made in North Carolina. Wood in another form takes the place of wool in warm winter clothes. Wood that has been specially treated in one way is better for some uses than rags of cotton cloth. And wood specially treated in another way takes the place of metals in certain kinds of construction.

The waste products of pulping may turn into important and valuable by-products: a substitute for vanilla flavoring, a synthetic tanning extract, a water softener, a substitute for hard rubber in storage batteries.

Perhaps the state's greatest opportunity for expansion lies in chemical industries. North Carolina was still mainly agricultural when some of the other states were

becoming industrial. By the time industry in North Carolina started to develop, these other competing industries were already well established.

In chemical industries North Carolina has an equal chance with other states to start at the beginning. Workers have a chance to grow as the industry grows, to learn special skills as the industry develops the need for them.

So much for some of the industries North Carolina needs in order to have a better balance between resources and use.

How can these changes be made? It is not an easy job. It is probably more difficult to do this than to carry out plans for better agriculture, and it may take longer, too. In farming each man may be owner, manager, and worker of his industry all in one. If he decides to plant different crops or to follow different cultural methods, he can do so. And there are state and federal agencies to help him.

But the worker in manufacturing, no matter how much he may want a better-paid job, has to wait until the job is available. Or he must go somewhere else to find it. Because this is true, in some years thousands of workers have left North Carolina to get factory jobs in other states.

If a worker wants to keep on working in North Carolina, he may have to keep a low-paid job and wait until industries with better jobs are established. He cannot start a new industry all by himself. At least, not directly.

Through the state government, people in North Carolina have set up special agencies to develop all the resources of the state. The North Carolina State Planning Board worked toward better balance between agriculture and industry. The Board's reports show the past development of agriculture and industry, and point out the needs and possibilities for the future. This Board was abolished in 1947.

The North Carolina State Department of Conservation and Development, which was set up to conserve and develop our resources, does its job in many ways. Among other things, the Department advertises North Carolina's advantages—in location and climate, in labor and raw materials—for the establishment of new industries in the state.

At the State College for Agriculture and Engineering of the University of North Carolina, the Textile School carries on research with fibers, yarns, and fabrics, trying to find better methods and better products. The school, which often works with textile manufacturers on their special problems, receives financial aid from the North Carolina Association of Cotton Manufacturers.

The School of Engineering at State College is aided by the North Carolina Engineering Foundation, which was organized by men in business and industry who realized the importance of engineering education and research.

The Business Foundation of North Carolina helps to support the work of the School of Commerce of the University of North Carolina. Men in business and men in the University are joined together in a program for more research and education, for better business and better industry in the state.

North Carolina's business men have already begun to make manufacturing more diversified. They know that diversification will mean a more prosperous state.

And during 1946 some of our leading citizens started a state-wide program to develop rural industries. Scatter industries over the state, and bring industry to the people in smaller towns and villages and in the open country, these leaders say. Then workers will not have to move from their homes to cities, which become more crowded every year. And there will be markets close by.

If we can develop rural industries to serve local communities in every part of North Carolina, the state will be one of the pioneers in a new industrial movement.

Newspaper editorials and public speakers, radio programs and magazine articles have pointed the way toward diversified industry and toward rural industries. The people, young and old, in city and in country, must do the rest.

Now Is the Time!

To Check Up and Sum Up

1. Bring your word list up to date.

2. Make a two-column chart showing in one column what we have in the way of industries in North Carolina

and in another column what we need in order to have a balanced development of the state's resources.

3. Make a large industrial map of your county, locating by symbols the chief industries.

4. Imagine you are a young man who wants to start some new and promising business in your home county. What industry do you think you would try to start? What things should you take into consideration before deciding? From what people or agencies could you secure information and advice? How might you obtain the needed capital?

To Think Through and Discuss These Questions and Problems

1. Should North Carolina try to manufacture all the things she needs? Why or why not?

2. What are some industries which would probably not be profitable in North Carolina? Why do you think these would not be profitable?

3. What are the advantages and disadvantages of small industries? (You might even debate this question.)

4. In what way does diversified manufacturing depend upon capital wealth? Increase capital wealth?

5. In what way does it depend upon technological wealth? Increase such wealth?

6. In what way might it promote diversified farming?

7. In what way does it depend upon human wealth? Enrich human wealth?

8. In what way might it increase institutional wealth?

To Share Your Information and Ideas

1. Why not have another exhibit, this time to show the products of North Carolina manufacturing? After it is over, select for the permanent school museum, begun earlier, those items which will be of lasting interest.

2. Why not make a frieze or a mural showing the highlights in the industrial development of North Carolina?

The power of falling water gives North Carolina one of its great resources for manufacturing.

V. INCREASING OUR INSTITUTIONAL WEALTH

Who can vote in North Carolina? Anyone may vote who is a citizen of the United States, is twenty-one years old, can read the Constitution and can write, has lived in the state for one year and in the county and precinct for four months.

40. *Government by the People*

THE permanent white settlers of North Carolina—Englishmen, Scots, Scotch-Irish, Germans, and all the rest—came in search of a chance to make a better living. They brought with them whatever they needed to conquer the wilderness—simple skills and simple tools. Some of them, coming in search of freedom, also brought with them a belief that everyone has certain essential rights.

In order to make these rights clear, the colonists in North Carolina held their first Provincial Congress at New Bern in August of 1774. The people were planning no revolution, had no idea of separating from Britain. Delegates to the Congress drew up resolutions declaring "That His Majesty George the third is lawful and rightful King of Great Britain, and the dominions thereunto belonging, and of this province as part thereof, and that we do bear faithful and true allegiance unto him as our lawful sovereign. . . ." And they went on to say "That we claim no more than the rights of Englishmen. . . ."

But they thought that they were not being given their rights as Englishmen. They had a number of complaints against the government, and listed various reasons why they were dissatisfied with the way they were being treated.

When the Fourth Provincial Congress met at Halifax almost two years later, in April of 1776, feeling in North Carolina had changed. Instead of declaring allegiance to Britain, the resolutions of this meeting declared that "the King and Parliament of Great Britain have usurped a power over the persons and properties of the people unlimited and uncontrolled." The colonists were convinced that in order to have their rights, they must be independent.

North Carolina was the first colony to declare independence from the mother country, in the Halifax Resolves of April 12, 1776. On July 4 of the same year, representatives from all the colonies signed a joint Declaration of Independence. Then each colony set about making plans for a government.

Delegates elected by the people of North Carolina met again at Halifax in November of 1776 to write a constitution for the state. The first thing they did was to make a list of what they considered the essential rights of man. Some of these were rights which were a part of every Englishman's heritage. Others were new rights which gave people greater freedom—and greater responsibility—than they had ever had before.

The delegates declared that no freeman could be imprisoned except by the law of the land, that every freeman had a right to know what he was accused of, and that no freeman should be convicted of a crime except by the unanimous verdict of a jury "of good and lawful men."

These ideas are now so familiar that the words and phrases have lost some of their meaning; but to the men who met in Halifax in 1776 these rights were very real, very important, and the words were full of meaning.

TATE GOVERNMENT IN NORTH CAROLINA

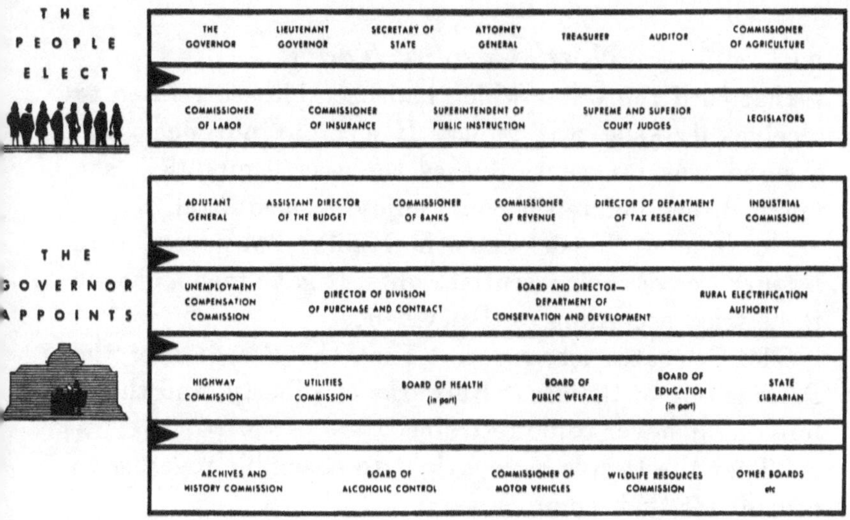

One of the most serious complaints of the colonists against Great Britain was that men were being thrown into jail without any warrants for their arrest. A writ of *habeas corpus* (a Latin phrase meaning "you may have the body") protected men from such illegal imprisonment by requiring a jailer to bring forward anyone being held in jail illegally.

Before the time when there was trial by jury, it was common procedure, and entirely legal, to try a man by ordeal.

In the ordeal by fire, an accused person had to walk barefoot over burning coals or red-hot plowshares, or he had to dip his hand in boiling water or boiling oil or melted lead. If the skin showed no sign of being burned, the accused was declared innocent.

There was also ordeal by water, which was used especially in the trial of women accused of being witches. The accused was first tied, and was then thrown into a lake or other water which had been blessed. If she floated

—that is, if the water which had been blessed refused to receive her—she was guilty. If she did not float, the accused was innocent. But as she was firmly tied, she might drown before she could enjoy her acquittal.

No wonder the delegates in Halifax considered trial by jury one of their essential rights. It gave men a chance to be tried according to the evidence.

The delegates also wrote "That the Freedom of the Press is one of the great Bulwarks of Liberty, and therefore ought never to be restrained.

"That the People have a right to Assemble together to consult for their common good. . . ."

Freedom of the press. Freedom of speech. Freedom of assembly. These phrases continue to mean something, because they are freedoms which are still being put to the test.

Groups of people often meet together to express ideas which are not the ideas of the majority, opinions which are opposed to the opinions of those in power. Sometimes these minority groups are concerned with matters of government, sometimes with matters of religion, sometimes with labor organization or other economic matters.

In times of peace people have the right to meet together and talk about their ideas, as long as the speakers at the meetings do not break the laws of decency and of libel. This right of the minorities is one of the great rights of a democratic government. It is one of the freedoms that keep our government democratic.

The delegates also promised freedom of religion: "All Men have a natural and unalienable Right to worship Almighty God according to the Dictates of their own Conscience."

This is another right which is generally taken for

GOVERNMENT OF A COUNTY IN NORTH CAROLINA

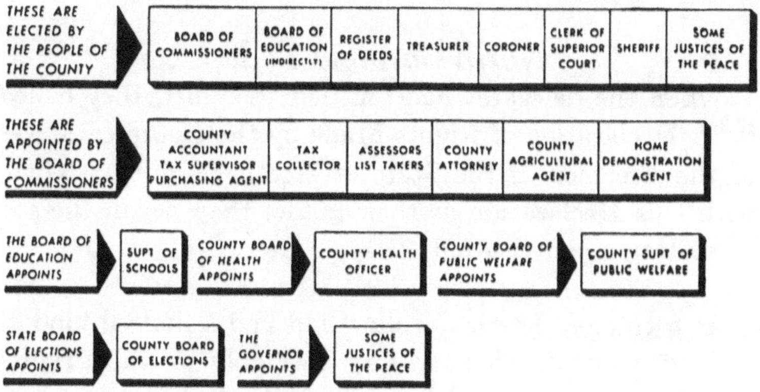

granted. Yet there are times when people forget that freedom of religion is guaranteed in North Carolina.

To these ideas about the rights of man, the delegates added certain ideas about government.

They thought the people should not be taxed without representation: "The People of this State ought not to be taxed or made subject to the Payment of any Impost or Duty, without the Consent of themselves or their Representatives in General Assembly freely given."

They divided the duties of government among three branches, so that no single branch would have all the power: "The Legislative, Executive and Supreme Judicial Powers of Government ought to be forever separate and distinct from each other."

They separated the church from the state: "There shall be no Establishment of any one Religious Church in this State in Preference to any other; neither shall any Person, on any Pretence whatsoever, be compelled to attend any Place of Worship, contrary to his own Faith or Judgment. . . ."

And they stated a fundamental principle of democracy: "All political Power is vested in and derived from the People only."

When the delegates had finished their list, they called it "A Declaration of Rights Made by the Representatives of the Freemen of the State of North Carolina." Then, with this Declaration as their guide, they began the job of writing a constitution.

As a colony, North Carolina had had a limited kind of self-government. There was an Assembly to which representatives were elected by the people. But the power of this Assembly was small compared to that of the Governor, who in those days was appointed by the King and governed with the authority of the Crown. Colonial government was more a government *of* the people than *by* the people. And government *for* the people as we know it today was to begin much later.

Under the North Carolina Constitution of 1776 there was a General Assembly made up of two houses, a Senate and a House of Commons. This General Assembly had a great deal of power. It not only passed all laws. It elected the governor, and it appointed judges and an attorney-general "who shall be commissioned by the Governor and hold their Offices during good Behavior." The Assembly also elected "a Council of State for one Year; who shall advise the Governor in the Execution of his Office." Even the limited duties of the governor were sometimes assigned to him "for the Time being" or "with the advice of the Council of State."

How was this powerful Assembly chosen? Since the Declaration of Rights had said that political power came from the people, we would expect to find that the Assembly was elected by the people. And so it was—but only by some of the people, not by all of them. Of course slaves and indentured servants could not vote. And

GOVERNMENT OF A CITY IN NORTH CAROLINA

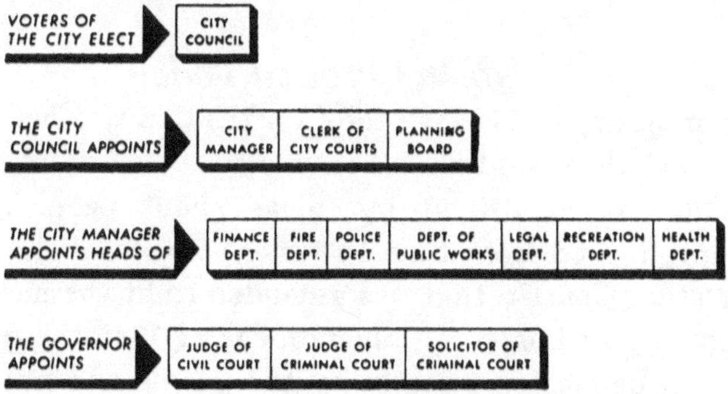

women could not vote. But not even free men could vote unless they met certain requirements.

Any freeman twenty-one years old, white or colored—if he had lived in the state for twelve months before election and had paid his taxes—could vote for a member of the House of Commons from his county. Any freeman twenty-one years old, white or colored, could vote for a member of the Senate if he had lived in one county for twelve months before election and had owned fifty acres of land for six months before election. The North Carolina Senate was elected by men of property.

There were also property requirements before a man could run for election to the General Assembly—requirements which made the power of landowners even greater. Each member of the House of Commons had to own at least a hundred acres of land. Each member of the Senate had to own at least three hundred acres. And before a man could be elected governor, he had to own property "above the Value of one Thousand Pounds."

Among the men who met at Halifax there were some who were in favor of making the Constitution more democratic than this. They believed that the governor should be elected by the people and that any free man should be allowed to vote or hold office, whether he owned property or not. The delegates with this opinion were not in the majority in 1776; so the Constitution was not written the way they wanted.

But as years went by, ideas about people's rights changed. And from time to time, by vote of the people, the state Constitution was amended to fit these changing ideas. According to an amendment of 1835 the governor was to be elected directly by the people and not by the General Assembly, and the right to vote was taken away from colored men. In 1857 men no longer had to own land in order to vote or hold office. After the War between the States, Negroes had the right to vote. In 1920, in a federal amendment passed by the states, women were given a nation-wide voice in the government.

Ideas about the organization and the duties of our government have also changed. The population of the state increased, and small villages grew into towns and cities. People living close together in these communities established their own governments to take care of local needs and to work with the governments of county, state, and nation.

Our county governments are now called on to do a number of different jobs, many more than counties were expected to do in the early days of the state. County officers often help to bring state and sometimes federal laws and services within reach of all our citizens.

Our state government, too, is now a complicated network of agencies. North Carolina citizens, through their representatives in the Assembly, have voted to do many things for themselves. In the state government are some of the agencies which carry out the will of the people.

On page 313 of this book there is a chart which outlines the organization of our state government. And on pages 315 and 317 are other charts showing the organization of one of the counties and one of the cities in North Carolina. Our city and county governments are not all the same, but they follow these general plans. These charts make it easy to see how complicated our government has become since men met in Halifax in 1776 to write North Carolina's first constitution.

You'll Want to Read More About the Struggle for Government by the People in North Carolina

1. See what your history and reference books say.

2. Read and report on such books or stories as these:
   a. Boyd, *Drums*—a novel about colonial and revolutionary times.
   b. Connor, *Makers of North Carolina History.*
   c. Stillman, *Drums Beat in Old Carolina*—a thrilling story of North Carolina during the days of the Regulation.

You'll Want to Know More About the Government of North Carolina Today

1. Get copies of the present constitution. Read and discuss the Declaration of Rights (Article I) and compare it with the Declaration of Rights in the Constitution of 1776. Free copies of the Constitution may be secured from the State Department of Public Instruction in Raleigh.

2. Study the charts shown in this chapter. See what the Constitution says about the organization and powers of state, county, and city governments. The State Department of Public Instruction will send you free a wall chart showing the organization of the state government.

3. See Warren, *The People Govern North Carolina*, Chapters 2–5, 9–10, 18–19.

4. Learn what the present requirements for voting in North Carolina are.

5. Find out the name of the state representative from your county. Do you live in a county with more than one representative? Who is the state senator from your district? What are some of the abilities you would look for if you were voting for members of the State Assembly?

You'll Want to Think About These Questions

1. How is government by the people endangered when good citizens do not take the time or the trouble to vote in elections?

2. What would you say if you were going to make a radio speech the day before an election in order to get people to go to the polls to vote?

3. How does the division of government power and duties help safeguard our rights as citizens?

41. *Government for the People*

AT FIRST, government meant the laws describing what people in a community could and could not do, and what they should and should not do. In a way these laws are like the rules for playing a game. Teams that play games together agree to obey certain rules. If they break these rules, they must pay certain penalties. People in a community or a state or a nation agree to obey certain rules, or laws. Anyone who breaks the law must pay the penalty.

For a while it was the chief duty of government to make laws under which numbers of people could live and work together without interfering with the rights promised to everyone by the Declaration of Rights and the Constitution of North Carolina.

But people gradually came to want other things besides these rights. They wanted their children to have an education. They thought their work should pay them enough to live comfortably. They wanted protection from disease.

Those who were unable to take care of themselves should be taken care of. There should be public parks and other places of recreation where people might have the pleasures and benefits of North Carolina's natural playgrounds. They wanted paintings to look at, and music to listen to, and books to read.

Through the representatives they send to the General Assembly, citizens of North Carolina decide how they will spend the tax money for state government services.

People saw that if they worked together through their government they might do what would be impossible for one person, working alone. So by vote of their representatives the citizens of North Carolina decided to use part of the tax money of the state to get things for themselves and their children.

The state could not afford to do everything at once. There was not enough money. The people began with the University. Then they opened free public schools. When there was more money, public health and public welfare were added. Later, public roads were built. As people made more money and paid more taxes, these agencies were able to do more. From time to time, citizens voted to spend tax money for some of the other things they wanted.

Every two years, when the General Assembly regularly meets, one of the important duties of the people's representatives is to decide how the state's money should be spent during the next two-year period.

Some of the money is used for the general expenses of running the government. There must be salaries for members of the Assembly, for the Governor and other executive officers, for the officials of the state courts, for everyone employed in state offices. All the costs of administration have to be taken care of.

The Assembly also decides how much of the state's income should be spent on the various agencies which have been established for the common good. If we look through a partial list of our state appropriations for one year, we get a general idea of what the people of North Carolina are doing for themselves.

The list on page 324 shows the amounts set aside for many special state services. In addition to these, our tax money must also pay for the general expenses of government. These include amounts for the Supreme and the Superior Courts, for the Governor's office and all the other offices of administration, and for the next meeting of the General Assembly.

HOW THE PEOPLE OF NORTH CAROLINA PLANNED TO SPEND SOME OF THE TAX MONEY IN 1947-1948

For Agricultural Improvement	$2,742,501
For Industrial Improvement	349,598
For Highways and Public Works	62,937,958
For Hospitals and Dept. of Health	7,932,452
For People's Aid and Welfare	10,098,049
For the State's Colleges & University	5,045,590
For the Public Schools	63,730,954
For Libraries	323,440
For Dept. of Archives & History	53,821
For Dept. of Conservation & Development	1,069,947
For the State Art Society	5,000
For the North Carolina Symphony Orchestra	12,000

The agencies paid for by public money for public good are important in the life in the state. Along with other schools, libraries, colleges, universities, hospitals, clinics, laboratories, art collections, and orchestras they make up part of the institutional wealth.

The government of North Carolina has become more than a government of the people and by the people. Today it is a government *for* the people, doing things for the public good. Government now has two main jobs. It passes, and enforces, the laws which are necessary for community life. And it carries on the services which the people of the state have established for themselves.

Have You Ever Thought About These Questions?

1. What is the difference between government *by* the people and government *for* the people? What is the connection between them?

2. How much do the people spend each year for state government and government services? Where does the money come from?

3. What services performed by the state cost the most?

Why Not

1. Read carefully the list of appropriations on page 324?
   a. What surprises you most about the appropriations?
   b. Which appropriations affect you most directly?

2. See North Carolina Manual for list of state agencies and boards?

3. Consult these references to find out about taxes levied by the state?
   a. Arnett, *The Story of North Carolina*, pp. 389–394, 438–440, 471–474.
   b. Newsome and Lefler, *The Growth of North Carolina*, pp. 432–434.
   c. Warren, *The People Govern North Carolina*, Chapters 6–8, 24.

4. Interview local officials to get information about the money collected and spent by your county and town?

42. *Schools, Colleges, and Universities*

WE GENERALLY take our public schools for granted today, but we have not always had them. So to measure North Carolina's institutional wealth in schools, perhaps we should know something of the way they developed.

How much schooling a child could get in colonial North Carolina depended on how much money the family had. There were no free public schools.

Children of the well-to-do generally learned to read and write at home. They were taught by a member of the family or by a tutor or a governess. If the family lived far away from any community, the children might have a private teacher until they were grown.

In towns the sons and sometimes the daughters of well-to-do families often attended private schools. In addition to reading, writing, arithmetic, and grammar, the girls might learn to embroider, recite poetry, and play the piano. This was as much schooling as girls received. There were no colleges for women.

Boys from wealthy families might go on from private school to college in Virginia or to Harvard, Yale, or Princeton in the North; and some were sent to England to attend the university at Oxford or Cambridge.

Education for poorer people was limited. The children

might learn to read and write and to work simple problems in arithmetic—probably not very well, because they were taught by mothers who often knew very little themselves. Without free public schools poorer children had little chance to learn. They started working while they were still young, and when they grew up and had children of their own, they could teach them only what they had been taught by their own parents.

Poorer people in general were not at all disturbed because their children could not go to school. For years education had been meant for the rich, not for the poor. "What good would it do, even if the children went to school?" they asked. "Schooling wouldn't help anyone get a better job or more pay."

But toward the end of the eighteenth century, ideas about education began to change, along with ideas about government and the rights of the individual.

Some of the delegates to the North Carolina Constitutional Convention in Halifax still believed that education should be kept for the few, just as some of them thought that a man should not vote or hold office unless he owned property.

Yet in their Declaration of Rights the delegates wrote, "All political Power is vested in and derived from the People only." If this is so, some of the men said, then all free men must be able to vote. And this means that all the people must be educated. They must be trained to understand and think.

So in the Constitution of 1776 there is a section which begins like this: ". . . a School or Schools shall be established by the Legislature for the convenient Instruction of Youth . . . and all useful Learning shall be duly encouraged and promoted in one or more Universities."

Good roads and school buses make it easy for children living in the country to attend school regularly.

These words in the Constitution were mere words, and nothing more. Probably only a few people really believed in the idea which these words expressed. Most of them had no interest in public education. Many of them, the poor as well as the rich, were against it. Years passed before people came to believe in it and work for it, and by that time North Carolina was far behind many of the other states.

Nothing at all was done about starting free public schools. North Carolina was not yet ready to pay for education by raising money through taxes. Free schools must wait.

Early governors and other leaders in the state kept asking the Assembly to start schools. In 1817 Archibald D. Murphey urged North Carolina to do something about educating children—that is, the white children. There were no plans then to have schools for Negroes. In fact, in 1830 a law was passed making it illegal to teach slaves to read or write, except figures.

In 1835 the North Carolina Constitution was revised and made more democratic. The governor was to be elected directly by the voters, and not by the Assembly. Representatives in the House of Commons were divided among the counties according to population. This meant that ordinary people had more voice in the government.

Perhaps this is one reason why free schools were finally started. The poorer people, listening to all the talk about the importance of education, may have supported a public-school law because it gave them their only chance at any schooling. And the wealthier people may have believed that a man who votes should not be ignorant.

Later, when a man no longer had to own property in order to vote or hold office and there were more people voting, schools for the poor were even more important. With growing democracy in government, the need for education became clearer.

A law for public schools was passed in 1839. In 1840 the first schools were opened. Ten years later there were over 2,500 schools with more than 100,000 pupils. By 1860 North Carolina had a good public-school system. It looked as if the friends of education had won their fight.

But after the War between the States the schools were closed. When they finally opened again, there was very little money to pay expenses, and few parents seemed to care whether or not their children had any education.

After a while, with growing industries and better business, the state became more prosperous, and could afford better schools. But many people seemed to think low taxes were more important than education. So the Assembly did not make the taxes large enough to pay for better schools.

Before the War between the States, schools had had a four-month term. By 1890 some of them were open only two months, and less than half the children in the state went to school. Thousands of people, more than a third of all those ten years old and older, could not even read and write.

Public men who believed in the importance of education were worried, and tried to get the people to support better schools. In 1889 the State Board of Education made plans for teachers' institutes all over the state. Speakers traveled from one end of North Carolina to the other, holding meetings and talking to the people in every county. Newspapers used their columns to stir up interest. When Charles B. Aycock was running for governor, he promised schools for everyone—white and Negro—if he were elected. Under his leadership as governor, the state began to have better schools.

The counties raised money for schools by local taxes, and the General Assembly set aside part of the state's money so that schools in poorer counties could be better. This money was called the Equalizing Fund because it helped to make the schools more equal.

While Aycock was governor, hundreds of schoolhouses

Teachers and students need well-equipped laboratories.

were built, more than a thousand in all. And after his term of office was over, building went on. In colleges, normal schools, and special institutes North Carolina teachers were given better training for their jobs than they had ever had before. The state helped establish libraries in rural communities. The legislature passed a law requiring all children between eight and twelve to go to school at least four months a year, and another law making it illegal for children under twelve to work. Then the regular school term was increased to six months. Education was on the march.

And all the time, more schoolhouses had been built—schools for white children, Negroes, and Indians; schools in remote mountain coves and far-away places along the coast, schools scattered throughout the state in every county. Without cars or good roads, children had to walk to school; so schoolhouses were built within walking distance of rural children. Many of the buildings were small, often with only one room and one teacher for all grades. But more and more children were having a chance to go to school.

Then North Carolina started building better state and county roads, and the highways which meant so much in the development of agriculture, industry, and business played a big part in the development of the schools, too. Children rode to school over these roads in buses operated at public expense. Attendance improved when students did not have to stay home because of cold or rainy weather.

When it was no longer necessary to have schools within walking distance, larger consolidated schools were built, and small neighborhood schoolhouses were closed. And though a larger school is not better than a smaller one just because of its size, it has a chance to be better. It has more classrooms, more teachers for different grades and different subjects, and probably better equipment.

A consolidated school, with students coming from a distance, draws a large community together. The boys and girls and their mothers and fathers meet more people, learn new things from each other. They share books and hobbies, skills and talents. More people work together for a better school.

North Carolina has come a long way since Aycock and other leaders started their campaign for better schools. Measured by the special yardsticks generally used, the state school system is much better than it used to be. There are twelve grades, the term is nine months long, and the compulsory age limit is sixteen years. Textbooks in the first eight grades are free to all children—white, Negro, Indian, and everyone else. In other grades, textbooks are rented to the students.

Every year we are making our public schools better. Every year this part of our institutional wealth becomes greater. The state's public schools are here, ready for our young people to put them to better use.

The teachers in our schools are better trained than they used to be. In 1945–1946 there were 22,808 white teachers being paid by the state. Of this number, 19,106 had had four years or more of college study. At the same time there were 6,851 Negro teachers being paid by the state. Of this number, 6,282 had had four years or more of college study. North Carolina can be proud of this record.

In spite of these changes for the better, North Carolina does not always compare well with other states. In 1940

Libraries must have plenty of well-chosen books.

the yardsticks showed that among the eleven southeastern states North Carolina ranked third, but among all the states in the nation North Carolina ranked thirty-ninth. After 1940 we did many things to make our schools better. But with all the improvement that has been made, there is still much to be done.

People are not satisfied just because the schools are better than they used to be. They are still working toward the ideal set for them by the men who met in the little town of Halifax in the early winter of 1776. This ideal has been gradually changing through the years, as the wealth of the state has grown. What looked like the goal in 1776 or 1830 or 1900 has been reached and passed, and still people go on planning and working for better schools. North Carolina wants the best schools that can be had "for the convenient Instruction of Youth. . . ."

And what of the university where all useful learning was to be encouraged and promoted? In 1789, thirteen years after the Constitution was written, the University of North Carolina was chartered by the Assembly. It was the first state university to open its doors to students.

When the University was chartered, the Assembly made no plans to raise funds for buildings or for faculty salaries. Each person who gave the University a certain amount of money was promised free tuition for one student. The six people who gave the most could have University buildings named after them.

For a number of years the University of North Carolina was the only university in the state. And there were no colleges. Many of the young men who became leaders in North Carolina were educated in Chapel Hill.

Several colleges were founded by church groups interested in training their ministers as well as in education in general. In 1834 Wake Forest Institute was established by the Baptists. This later became Wake Forest College. In 1941 the college opened a medical school in Winston-Salem with funds given by the Bowman Gray Foundation. In 1946 Wake Forest received additional funds from the Reynolds Foundation, and started making plans to move to Winston-Salem.

In 1837 Davidson College was opened by the Presbyterians, and Guilford College was opened by the Society of Friends. In 1838 Union Institute in Randolph County was opened by the Methodists. This school changed its name to Normal College and then to Trinity College. It later moved to Durham, received a large endowment from James B. Duke, and became Duke University. It is no longer under the control of the church.

The state also established more colleges. In 1889 North Carolina opened the State College of Agricultural and Mechanical Arts in Raleigh. This is now the State College of Agriculture and Engineering, and part of the University of North Carolina. In 1889 the State Normal and Industrial College for training teachers was opened in Greensboro. This is now Woman's College, also a part of the University.

In 1945 the Educational Directory of the United States

Are the young people of North Carolina using the state's wealth in colleges and universities?

Office of Education shows that North Carolina had fifty-two institutions of higher learning. There were twenty-five colleges and universities, seventeen for white students and eight for Negroes. There were seven state-operated colleges for the training of teachers—three for white students, three for Negro students, and one for Indian students. There were twenty junior colleges, nineteen for white students and one for Negro students.

All these colleges and universities add to the institutional wealth of North Carolina, whether they are supported and controlled by the state, by churches, or by private corporations. How much they add depends on how good they are.

Colleges and universities have made their own yardsticks by which they measure and grade themselves. There is a Southern Association of Colleges and Secondary Schools, with certain requirements for membership.

And for the whole nation there is an Association of American Universities, with even higher requirements.

What is the measure of North Carolina colleges and universities, judged by these standards? In 1945 fourteen of the colleges and universities were approved by the Southern Association, seven were approved by both the Southern Association and the Association of American Universities. Four were not approved by either association.

Of the seven colleges for training teachers, three for white students and the two for Negroes were approved. Of the twenty junior colleges, five for white students were approved.

North Carolina has institutional wealth in its colleges and universities. We need to use this wealth, and keep on adding to it. If our colleges and universities are to mean all that they should in the life of the state, they must have the interest and support of the people of North Carolina.

Here Are Some References You'll Want to Read and Report on

1. Your history texts.

2. Gray, *Jane Hope,* a novel about life in Chapel Hill just before the Civil War.

3. Lathrop, *Educate a Woman*—a picture story of Woman's College, University of North Carolina.

4. Rowe, *Discovering North Carolina,* Chapter 14.

5. Salley, *Life at St. Mary's.*

Here Are Some Interesting Things to Do

1. Make a detailed study of one educational leader in your community or in the state. Write a theme or a dialogue which combines in an interesting way the information which you have gathered from many sources.

2. Represent in a chart the development of free public schools in North Carolina.

3. Try to find how much of the state's total income is spent for public schools. Represent this information in a graph.

4. Interview your school principal or superintendent to learn how much it costs to run your school and where the money comes from. Find out where the money for your school building came from.

Here Are Some Questions to Think About

1. Do you think taxpayers get their money's worth from what they spend on schools? Why or why not? What can you and your classmates do to insure them the greatest possible returns for their money?

2. What do you think should be our next steps in improving the educational opportunities of the people in North Carolina?

43. *Education of the People*

WHEN the University of North Carolina opened, students in Chapel Hill began studying subjects very much like those in any other university in the country. In the early public schools the courses of study were like those of the private schools. When the smaller colleges opened their doors, students there had about the same kind of courses as students in similar colleges anywhere else.

The elementary schools generally taught reading, writing, arithmetic, and geography. In the colleges and universities everyone received a general classical education, with courses in Latin, Greek, and English literature, history and political philosophy, astronomy, natural philosophy, mathematics, chemistry, and similar subjects.

For generations schooling had been limited to the children of well-to-do families. They were young men who would be lawyers or doctors or ministers or political leaders, not business men and certainly not men who worked with their hands.

When education became the right of everyone, people asked "What is the good of learning to read and write? What is the good of going to college? Or to the University? Does it get you a better job? Or more pay?"

These were natural questions, since many of the students would have to go to work, and would have to be able to work with their hands. They needed to know how to do things.

Many colleges began adding courses that were practical. The State College of Agricultural and Mechanical Arts in Raleigh was opened because some of the leaders in the state wanted the young men to have practical training. The State Normal and Industrial College in Greensboro was opened to train teachers for the public schools of the state.

Before the days of free public education, children started working as soon as they were old enough, instead of going to school. A boy worked with his father or as an apprentice with some other expert craftsman, learning the special skills that experience can teach better than anything else.

After children began going to school instead of working, schools started teaching some of the practical skills. Today students may learn in school the manual skills of working with various tools, the farming skills of better agricultural methods. There is training in bookkeeping, typing, and shorthand. Many household skills which girls used to learn by helping at home are now being taught in school: sewing, cooking, and canning and preserving foods.

Vocational training like this may be helpful to anyone, whether or not he expects to work in a job where these skills are needed. Some students learn skills which they will use later on in their work. But if a student spends a great deal of his time taking courses which give him vocational training, where will he have a chance to get a general liberal education?

This is one of the difficult questions in education today.

Students in our schools today may learn to work with their hands.

Everyone seems to need the same general liberal education, but everyone does not need the same vocational training. How can we give everyone some of the liberal education he needs and some of the vocational training he may find useful? This is a problem the schools are still trying to solve.

The people of North Carolina need more practical training. There is no question about that. They need to know how to do more things and how to do them well. They need more know-how, more technological skills. This was clearly seen in the study of North Carolina agriculture, which showed that many farmers in the state need to practice better methods of farming. And it was also seen in the study of North Carolina industry, which showed that North Carolina has many factory workers who are unskilled or semi-skilled and who are paid low wages because of this. Everyone in the state will make a better living when people are better trained and are able to earn higher wages.

More skills and better training have seemed so necessary that many people have been forgetting something else: the importance of a general, or liberal, education.

With a liberal education people are better able to understand themselves and the world they live in. By studying the past they are better able to see what the present means, better able to say what the future may or ought to be. Knowing facts, they can go beyond them.

Man is the only form of life able to do this. Man alone has a chance to profit by the mistakes of others, a chance not to make the same mistakes his grandfathers and great-grandfathers made before him. It is through such knowledge that man may go forward toward the kind of world he has been working for, generation after generation, century after century.

For a while North Carolina had to put buildings first on the list of what was necessary to make education better. Then practical training seemed important. Good equipment and good training are still necessary. Skills and know-how help produce the wealth that North Carolina must have in order to develop the other kinds of wealth. But now people realize that a general liberal education may give students the most valuable kind of training anyone can have.

People have come to understand that schooling and education are two different things. A person may be well educated without any formal education. Or he may have a college degree and never be educated at all. Education is never really finished with graduation from school or college or university. Education continues as long as a man lives. Through reading and experience his knowledge continues to grow.

Today people know that if their children do not go to school, they are missing one of the rights that citizens have been working to give them. School is one of the things people have planned for themselves through their representatives in the government. Yet even with free public schools there have always been many people in

North Carolina who were not literate. They never learned to read and write.

The census takers of 1900 asked everyone ten years old and older, "Do you know how to read and write?" And almost twenty out of every hundred white people in North Carolina and almost fifty out of every hundred Negroes answered "No." It is no wonder that leaders in the state realized something had to be done about the schools.

In 1930 the census takers asked the same question: "Do you know how to read and write?" This time North Carolina had a better record. Only a little more than five out of every hundred whites and a little more than twenty out of every hundred Negroes had to answer "No." In thirty years the schools had been able to do some of the things for which they had been planned and built. They had at least taught many people how to read and write.

In spite of this improvement, North Carolina still made a poor showing compared to other states. In 1930 only four states had more illiterates than North Carolina, and only one state had a larger number of native white illiterates.

Perhaps in 1940 the census would have shown greater improvement if the same question had been asked. But in that year the census takers began asking a new question. Men interested in measuring the literacy of the people decided that the old question did not always give a very good idea of how much people might or might not know. After all, a person who can read and write may not be able to read and write anything but his own name.

So in 1940 the census takers asked a new question of everyone who was twenty-five years old and older: "What was the highest grade in school you completed?" In the whole United States almost fourteen out of every hundred people in this age group, white and Negro, had not gone beyond the fourth grade. In North Carolina almost twenty-seven out of every hundred people in this age group, white and Negro, had not gone beyond the fourth grade. North Carolina's rank among the states was forty-second. Forty-one other states in the nation made a better showing.

The job of bringing schools to the children of North Carolina has been done. But there is still the job of bringing the children to the schools.

The state is doing this, too. The legislature has passed one law requiring children to go to school until they are sixteen. It has passed another law forbidding them to work before they are sixteen, except with special permission and in certain kinds of work. State buses make attendance better, and when children go to school regularly they have a chance to learn more. The age limit for compulsory attendance and for working means that children will be finishing more grades in school.

Many students want training in art and music and literature.

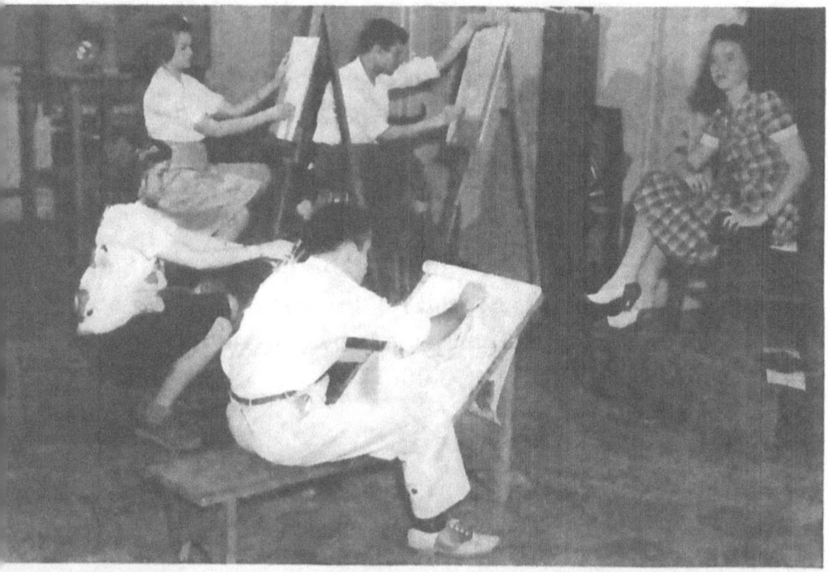

Things have changed since the day in 1839 when the General Assembly passed the bill for the first free public schools. Before then, school was the privilege of only the well-to-do. Afterwards, it became the right of any child. Now, with laws for compulsory attendance, school is a "must" for all children.

North Carolina is realizing that no state can be better than its citizens. To have good government, men and women must be truly educated. They must be informed and intelligent.

Besides starting a drive to improve the schools, leaders of the early 1900's also tried to do something about libraries in the state.

People who like to read and study want plenty of books, magazines, and newspapers. Yet they cannot always afford to buy all they want. Here again men found that by working together they could have some of the things that would be impossible for each man living and planning for himself alone. They established public libraries from which anyone may borrow books, where anyone may go to read and study. These libraries are one of the great agencies of the people.

In 1897 the city of Durham opened the first free public library in North Carolina. By 1910 there were 39 public libraries in various parts of the state, and by 1921 the number had increased to 67, three of which were especially for colored people.

In 1944 there were 68 county public libraries, 18 regional public libraries, 24 public libraries in cities and towns, and 35 public libraries for the use of Negroes. There were more than 1,200,000 books in the public libraries of the state.

Our libraries bring books within reach of thousands of North Carolina citizens.

Nine counties had no public libraries at all, and nine counties had only local or city libraries. The other 82 counties had libraries which were supported partly by money from the counties themselves and partly by money set aside by the Assembly to help the libraries of the state.

In 1944 the amount of money spent for public libraries in North Carolina averaged only twenty-one cents for each person in the state. This was exactly half the average amount spent for public libraries in the nation.

The North Carolina Library Commission, which was established by the General Assembly in 1909, sends books, package libraries, and traveling libraries to many people who have no public library near them. There are over forty bookmobiles which go into many parts of the state. Specially-built trucks, loaded with books on many dif-

A crowd is often waiting when the bookmobile stops. Young and old gather around to borrow something to read.

ferent subjects, now travel over county roads to people who may have no other way of finding what they want to read. On scheduled days the truck arrives, and a crowd soon gathers. Young and old have come to borrow books.

People living near colleges and universities may borrow books from their libraries. And the Extension Library of the University of North Carolina sends books by mail from Chapel Hill to people in every part of the state.

In 1901 the General Assembly passed the first law for rural school libraries. Since then, libraries have been added to public schools as an important part of the school program. The money for these libraries is included in the funds which the Assembly sets aside for the schools.

Our school libraries had more than 3,200,000 books in 1946, books on many different subjects and chosen especially for readers between the ages of six and eighteen. These books are within reach of more than 700,000 young people in the state. The libraries in our public schools bring books to many communities without any public libraries of their own.

Be Sure to Learn the Origin and the Meaning of These Terms

education, liberal education, illiterate, illiteracy

Be Sure to Do Some of These Things

1. Discuss the difference between schooling and education.

2. Find out:
   a. What library services your county has.
   b. How your school library ranks according to state and national standards.
   c. What agency or agencies would lend your class or teacher the books which you cannot secure in your own community.

3. Make a study of the courses now offered in your school. List them. Interview your parents and some of the oldest citizens to find out how the courses are different from those in the schools they attended. How would you account for the difference? Which courses in your school are meant to develop special skills, and which to give a liberal education?

4. Summarize in outline or paragraph form the chief facts in this chapter.

44. *Churches*

MANY of the people who settled in Carolina came from European countries where they had no freedom of religion. In England, for instance, there was an Established Church which was closely connected with the government and was supported by the government. The King of England was head of the Church as well as of the government. Everyone had to give money to support the Church of England, and for a long time it was against the law to belong to any other church.

Although settlers in North Carolina were free to choose what church they went to, there was also an Established Church in the colony. The Church of England was the Established Church, and the people were required to pay taxes to support it even when they belonged to another church.

Because many of the people in North Carolina had not always had religious freedom, they knew its value. So when the Declaration of Rights was written in Halifax in 1776, one section stated "That all Men have a natural and unalienable right to worship Almighty God according to the Dictates of their own Conscience."

Freedom of religion was promised as one of the rights of man. The North Carolina Constitution of 1776 also declared:

This church was built in Edenton in 1736, when coastal towns were the centers of government and trade.

> There shall be no Establishment of any one Religious Church in this State in Preference to any other; neither shall any Person, on any Pretence whatsoever, be compelled to attend any Place of Worship, contrary to his own Faith or Judgment; nor be obliged to pay for the Purchase of any Glebe, or the building of any House of Worship, or for the Maintenance of any Minister or Ministry, contrary to what he believes right. . . .

The religious freedom promised in the Declaration of Rights was made broader and surer by the Constitution.

After the Revolution the Established Church became the Protestant Episcopal Church. And because colonial North Carolina had religious freedom, there were already several other churches here. Among the very earliest settlers were Quakers—people who belonged to the Society of Friends. Most of the Scots and the Scotch-Irish were Presbyterians. Some of the Germans were Moravians. Many were Lutherans. There were probably Baptists among the early settlers, although a Baptist congregation was not organized until 1727. The Methodists had started in England in 1729 as a special group within the Established Church, but they were not active in North Carolina until just before the Revolution.

The Established Church had been the church of town and village. It was built where courts were held and government was carried on, where people came to buy and sell goods, where wealthy planters could gather for church services. The Protestant Episcopal Church continued in much the same way. It did not often reach out into the country. It remained in the towns. In 1860 there were only fifty-three Episcopal churches in the state.

Presbyterian churches were also built in towns and villages, especially in the neighborhoods settled by the Scotch-Irish and the Highlanders, in the rolling hills of the Piedmont and the lowlands of Cape Fear around Wilmington. By 1860 North Carolina had 184 Presbyterian churches.

Scattered throughout the state were many people who seldom had a chance to go to town for church services. Living on the farms they had cleared in the wilderness, they were far from any settlement, far even from their nearest neighbors.

Baptist preachers were among the first ministers who

North Carolina has many rural churches where members gather for worship. The country church is often the only religious or community center within miles.

traveled through the countryside to reach these scattered families. After a while, Baptist churches were dotted over the state, often standing at a crossroads without a house in sight but within walking or riding distance of rural families near by. By 1860 there were 780 Baptist churches in North Carolina. Of these, not more than thirty were in towns and villages. The others had been built in the country, for country people.

The Methodists, too, went out into the state to reach the people living in remote rural sections. At first there were no Methodist churches where congregations could meet for services. The ministers rode horseback over rough back-country roads, stopping to hold services for a family in the kitchen of some farmhouse or preaching to neighbors who gathered together in a camp meeting. Methodism became popular. By 1860 there were Methodist churches in every county, 966 churches in the whole state.

Besides Methodists, Baptists, Presbyterians, and Episcopalians, North Carolina in 1860 had congregations of Lutherans, Christians, Friends, Moravians, the German Reformed Church, and Roman Catholics. In all there were 2,117 separate churches, with 157,014 members—an average membership of a little more than 74 persons for each congregation.

Churches in the state have increased with the growth of North Carolina. In 1936 there were 8,029 different churches with 1,274,722 members. The average membership then was 159—only a little more than twice the average of 1860. Of all the churches in 1936, 1,893 were urban and 6,136 were rural. Of the members, 488,463 lived in towns and cities, and 786,259 lived in the country. North Carolina was still a rural state, and the churches were in all parts of the state, near the people.

Various Protestant denominations of the state have joined together as a group and formed the North Carolina Council of Churches. The Council of Churches states that its purpose is (1) to promote fellowship and better understanding among church members, (2) to advise on matters that concern the progress of Christianity in the state, and (3) to bring the churches together in the service of the Christian religion.

Through their churches people work to make the state better in many ways. Church groups were among the first to be interested in improving education in North Carolina. They founded several colleges so that young men could be trained as ministers. Before 1840 colleges for men had been established by the Baptists, the Presbyterians, the Society of Friends, and the Methodists. Early academies for women were opened by the Moravians, the Episcopalians, and the Baptists. Today North

Carolina has a number of colleges which are supported by various churches, but they do not limit themselves to students who are members of their own church.

Before people realized that the welfare of everyone should be the responsibility of everyone, the churches alone were doing much of the welfare work in which the state now plays a large part. Church groups helped to establish homes for orphans and hospitals for people needing medical care. They helped to give care and training to the crippled and the blind, so that many of them were able to help themselves. They helped to look after the poor people in their communities. The churches still do much work of this kind.

Churches today are important in the spiritual, intellectual, and community life in North Carolina. As the

The city of Charlotte has one of North Carolina's beautiful modern churches.

needs of the people have changed, the churches have met these needs, adding more and more to the institutional wealth of the state.

It Will Be Interesting and Informing

1. To take a religious census of your class. Let each student write on an unsigned blank which church he attends or belongs to, whether he goes to Sunday School or not.

2. To list all the ways churches are trying to make your community and state better. Divide your class into as many committees as there are church denominations represented, and let each committee study one church group. You may need to interview the ministers to get some of this information.

3. To make a church map of your town or county. Are there any areas without churches near by? If so, why?

4. To get up-to-date information about churches and church membership in North Carolina.
   a. Consult the section about churches in the latest issue of *The World Almanac.*
   b. Secure if possible the latest copy of the *Yearbook of American Churches.*
   c. Write to the North Carolina Council of Churches, Durham, North Carolina, for information.

5. To think about and discuss these questions, or write themes in which you express your answers to them:
   a. How do churches help people who do not belong to them?
   b. How do churches help strengthen other institutions?
   c. How can churches cooperate with one another more?

45. *People's Aid for the People's Welfare*

EVER since colonial days there has been some kind of government care for the poor, the insane, and the criminal. Some of the taxes paid by the colonists in Carolina were set aside to help poor people, and certain men in each community were responsible for looking after them. Men and women unable to work were given food and shelter, and orphaned children were taken care of.

At first only a few people in the colony were unable to support themselves. Most of the early colonists could produce their own food and clothing. As the population of North Carolina grew, it became harder for each family to produce whatever was needed. The number of people requiring public help gradually increased.

Colonial Carolina had no public institutions for the people's care. The only public buildings in most communities were the county courthouse and the jail. Because there was no other place for them, the people who could not support themselves sometimes lived in the jail, where they were fed at public expense. Men and women were often crowded into the same room, and sometimes those who were mentally ill were chained.

After a while the counties began building special houses for the poor. Under the same roof in these poorhouses were men and women too old or too ill to work; the crippled; the blind; the deaf; women whose husbands had died and left them with young children to be cared for; people with subnormal minds; and orphaned children who had no living relatives to care for them.

Everyone needing help from the local government lived in the county poorhouse no matter what kind of help was needed. That was the only place the county had for them. Those who might be dangerous because they were insane and those who had contagious diseases were allowed to live with those who were healthy.

For many years county poorhouses were the only places in North Carolina where needy people could get government help. Then, in the middle of the nineteenth century, the first institution to care for the people was built. This was the State Hospital for the Insane at Raleigh.

After the War between the States, the Board of Public Charities was established to look after all the charitable and penal institutions in the state.

During the 1880's, various citizens of North Carolina saw that many of their neighbors needed different kinds of help and needed it for different reasons. Church groups and fraternal organizations began to establish orphanages for the children of members who had died leaving no one to care for their families. In time, some of these orphanages received money from the state treasury to help with their expenses. Those receiving state aid at present are the Oxford Orphanage, Junior Order Orphanage, Oxford Colored Orphanage, and Pythian Orphanage.

There is a state-supported school at Raleigh for white children who are blind, and another at Morganton for those who are deaf. There is a special state school near Raleigh for Negro children who are blind or deaf.

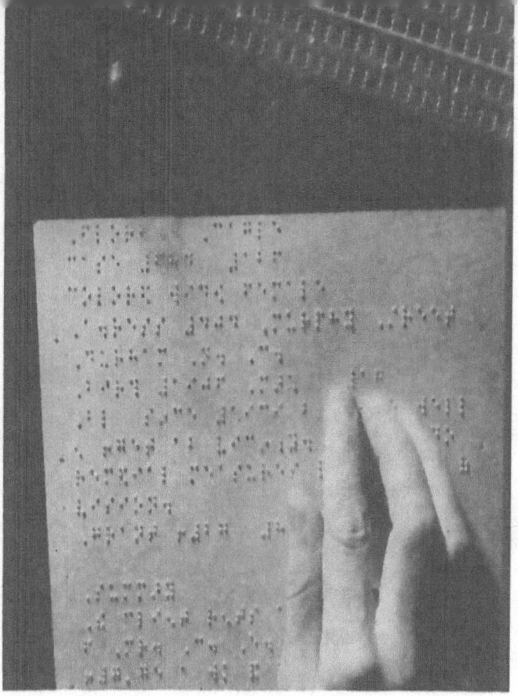

This is a page which most North Carolinians will never be able to read. It is written in Braille. Raised dots which stand for letters are made with the metal rule and pointed tool at the top of this picture. In schools for the blind, persons without sight are taught to read Braille with their fingers.

At Morganton there is a state hospital for insane white patients, and at Goldsboro there is a state hospital for insane Negro patients. Caswell Training School near Kinston cares for the white feeble-minded boys and girls, although it has room for less than half of the children needing this care.

In 1947 the General Assembly voted to buy one of the hospitals in the state which was owned by the federal government. Some of the people who are mentally ill are being cared for there.

Crippled children also have special help of their own. There is a hospital for medical and surgical care at Gastonia, known as the Orthopedic Hospital. And regular clinics are held in large centers over the state. In these clinics crippled children are examined and are sometimes given treatment. If necessary, they are sent to the Orthopedic Hospital for special care.

Three hospitals for tubercular patients provide care

for citizens with this disease: the State Sanatorium, the Eastern North Carolina Sanatorium, and the Western North Carolina Sanatorium. In many counties there are smaller hospitals, supported locally, for patients from sections of the state near by.

The state also looks after those who have broken the law and must be punished. There are special training schools for white boys at Concord and Rocky Mount; for white girls at Eagle Springs; for Negro boys near Rockingham; and for Negro girls at Rocky Mount. At these schools young people who have broken the law are given training in how to become responsible citizens. They are also taught to do various kinds of work, so that when they leave the schools they will be able to earn a living.

For adult offenders against the law, North Carolina built the Central Prison in Raleigh. There are also prison camps in various parts of the state. Most prisoners are now kept in these camps, and are sent out from there to build and repair the roads and highways.

In 1917 North Carolina's Board of Public Charities was reorganized as a state-wide agency, and was called the State Board of Charities and Public Welfare. In 1919 an act was passed requiring every county to organize for public welfare work. In 1945 the name of the state agency was changed to the State Board of Public Welfare. The Board now takes part in all county welfare work, and co-operates with state and federal welfare agencies.

The various agencies and institutions of North Carolina have done a great deal to give special help to the people. But in the 1930's things happened all over the nation and the world which made it necessary for the people, through their government, to plan new agencies and a different kind of help.

In the 1930's, when there was a depression, many who wanted work were unable to get it. They did not need the help which the government was giving the crippled or the deaf or the blind or the insane. These people were strong and healthy and able to work. But they could not find anything to do, and many of them did not have money for the simplest things essential to living—food, clothing, and a place to live.

This was true in all the states in the country, not in North Carolina alone. So many families needed help that the states were not able to care for them, and the federal government set up special agencies for the aid of the people. In 1935 over 25 per cent of the population of North Carolina was receiving government aid of one kind or another.

During the hard times of the 1930's the majority of the citizens of North Carolina and of the nation came to believe that some of these special agencies should be made a permanent part of the government. Studying the history of the past, they saw that there had been depressions ever so often. There were times when people who had jobs sometimes lost them, times when people could not always find jobs even when they were able and willing to work.

Yet in states like North Carolina, where the average income is low, it is not always easy to save anything for old age or for the time when this income may be even lower or for the time when there may be no work and no income at all. When people lose their jobs, they often have nothing to live on until times are better and they are once more able to find work.

In the 1930's new laws for the general welfare and the common good were passed by the federal and state governments. Under these laws help is given to people too old or too young to work and people unable to find work.

People who are sixty-five years old or older and who can show that they need help receive a small monthly income from a special fund. Half of the money comes from the federal government, a fourth from the state, and a fourth from the county. Old Age Assistance, as it is called, helps thousands of old people stay in their homes instead of going to a poorhouse as they used to do.

There is similar help for the care of blind people who need it, and this help is given to them no matter what their age is.

In the 1920's the state began giving help to mothers with young children who had no one to support them. In 1935 this state aid was continued under the National Social Security Act. Now federal, state, and county governments all contribute to the aid of mothers with dependent children. Instead of using money to care for these children in orphanages, the government agency sometimes pays the mother an income every week. She receives a certain amount of money for each child. This help often makes it possible for a mother to keep her home and have her children with her.

There are now special federal and state laws which provide help for workers in certain industries, laws which help the workers to help themselves. There is a fund managed by the Employment Security Commission. The money in this fund comes from some of the employers in the state. These employers pay into the fund 3 per cent of the amount they pay their workers as wages. Then, if through no fault of his own, one of these workers loses his

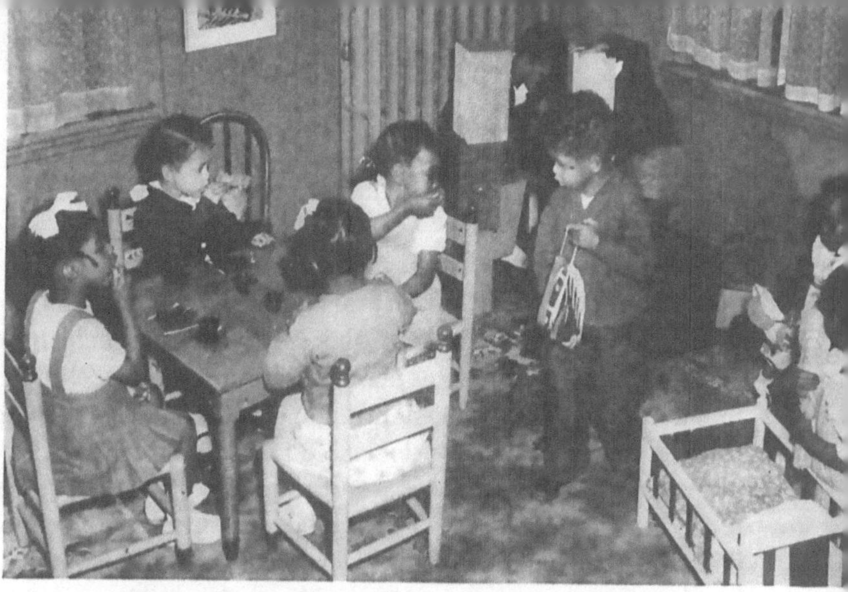

Some towns and cities have nurseries to take care of young children whose mothers are working during the day.

job, he can draw money from this fund. He receives a certain amount each week for a certain number of weeks while he is out of work.

Employers and workers both pay money into a fund for Old-Age and Survivors' Insurance. This fund is managed by the National Social Security Board. As soon as a worker is sixty-five years old, he may give up his job and live on a small monthly income from this fund. Or if he dies before he is sixty-five, his wife and dependent children receive the insurance money.

Certain standards of safety have been set up for workers in industry, and there is a workmen's compensation program administered by the State Industrial Commission. Insurance is paid into this fund on the basis of the worker's earnings. If the worker is injured at work through no fault of his own, this program provides money for medical and hospital expenses. And the worker can help take care of his family even when he cannot earn full wages because of his injury.

The one hundred County Departments of Public Welfare carry out many important services. Anyone can go to his County Department, where someone will help him get aid. Some people who need aid are not always eligible for help from state or general agencies. Perhaps they are too old to receive the special help which is given to children less than eighteen years old. And yet they may be too young for old-age assistance, which begins when people reach the age of sixty-five. These people are helped by the counties through their welfare offices. The County Departments of Public Welfare work in the local communities and are close to the people.

Besides helping people, welfare agencies now try to find out why help is needed, why people are poor or unable to get work. Why do they sometimes seem lazy? Why do people break the law? Is there any connection between crime and the way people live? What is the effect

State and private agencies give help to mothers with young children who have no one to support them.

on people when they live in small, poorly built houses, haven't enough to eat, and are often ill?

Welfare workers and other thoughtful citizens ask other questions. How much help can people receive without becoming so used to it that they take this help for granted? Will people learn to accept government aid as a matter of course, and expect the government to do more and more for them as time goes by?

The answers to these questions are not easy to find, and they are not always the same for everyone. What is true of one person may not be true of others. In the meantime, citizens are thinking about these problems. And welfare workers are trying to help people to help themselves.

The State Department of Public Welfare has white and Negro workers to aid the whites and the Negroes. There are county welfare agents. And the federal government sends money and workers to help with the work of state and county agents. These public agencies are all supported by taxes paid by the people. Many private agencies supported by churches, community groups, and clubs do the same kind of work. Community Chest funds now give help to many thousands of people in the state.

As capital wealth grows larger, North Carolina's institutional wealth increases, and the people are able to do more for the public welfare and the common good.

Have You Learned

1. What a public agency is?

2. Which public agencies give aid to each of these groups: the handicapped (the deaf, the blind, the crippled, etc.), the mentally sick, the aged, orphans, tu-

bercular patients, lawbreakers, the unemployed, and injured workmen?

3. What units or levels of government cooperate in giving such aid?

4. What welfare agencies do besides help those in need?

5. What private groups in your community help in giving aid to the needy?

What Do You Think?

1. Are there some groups needing help which are not being properly cared for by the state? If so, what are they? What should be done for them?

2. Is there a danger that people who receive aid from the government might come to do less and less for themselves? If so, what do you think can be done to overcome that danger?

3. What should be kept in mind by citizens when deciding what their government should do for them and what they as individuals or as private groups should do for themselves?

46. *A Healthier North Carolina*

HEALTH is an important part of the people's welfare, so important that North Carolina has a special Board which helps to look after the health of everyone in the state. In 1877, when the Board of Health was established, the General Assembly voted to give it $100 for its yearly expenses. In 1945 the Assembly voted to give the Board over $1,000,000 for the next two years. Instead of $100, the Board had more than $500,000 for its yearly expenses.

Besides the money voted by the people through the Assembly, other large amounts are being spent on health programs in North Carolina. Counting the funds from the United States Public Health Service, the Children's Bureau, and other organizations, and the money voted by local, county, city, and state governments, more than $5,000,000 was spent on better health in North Carolina in 1945.

Although the Board has never had money enough for all the work that is necessary, it does a great deal for the people in North Carolina. There are so many different things to be looked after that the Board now has a number of special divisions, and each division has a certain job for which it is responsible.

The Division of Vital Statistics keeps records of the

births and deaths in North Carolina, with the cause of every death.

The Division of Sanitary Engineering supervises general sanitation. It helps local health officers inspect dairies and plants where milk is pasteurized and prepared for sale. It inspects the water supply and sewerage systems of public schools and of towns and cities. It inspects restaurants, hotels, meat markets, and jails, to make sure that they are clean and that the food is being handled in a sanitary way.

The Division of Preventive Medicine aids local health departments in holding clinics for mothers and babies and for the crippled children in the state, and it pays surgeons and hospitals to care for them. It also supplies mothers of young children with special booklets on the feeding and general care of their families.

The Division of County Health Work advises with the public health officers in the counties and cities. Through these officers the state's program for better health is able to reach the people in town and country.

The Division of Epidemiology is concerned with the control of epidemics and of contagious diseases. It tries to give important information to everyone in the state through community programs on nutrition, nursing, and first aid and through special motion pictures about health. It has helped with many special campaigns for the control of disease. And it has cooperated with the United States Public Health Service in malaria control. North Carolina, like other states, sometimes has epidemics which spread rapidly from one community to another. Infantile paralysis and influenza are two such diseases. If there is danger of an epidemic, the Division of Epidemiology suggests ways to protect the health of the people.

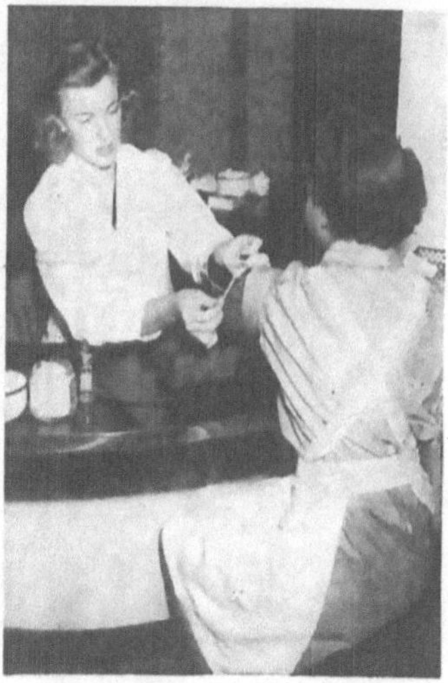

Workers handling food are given tests to make sure that they have no diseases which they might pass on to others.

The Division of Tuberculosis Control works toward the prevention and cure of this disease.

The State Laboratory of Hygiene at Raleigh is the oldest and largest division of the State Board of Health. The Laboratory carries out various health tests for the people, whenever it is asked to do so by physicians and local health officers. It examines the heads of dead animals to see if they had rabies. And it examines samples of drinking water sent in regularly by health officers and the officials of cities. From this Laboratory physicians and local health departments can get the vaccines and serums which help prevent certain contagious diseases, such as typhoid fever, smallpox, diphtheria, whooping cough, tetanus.

The Division of Oral Hygiene teaches care of the teeth and mouth to children, mothers' clubs, and study groups. It also arranges for dentists to examine the teeth of school children and to do any necessary dental work for those whose parents are unable to pay for this service.

The Division of Industrial Hygiene has developed as a result of modern manufacturing methods. It protects the health of people in industry. Some diseases may be the

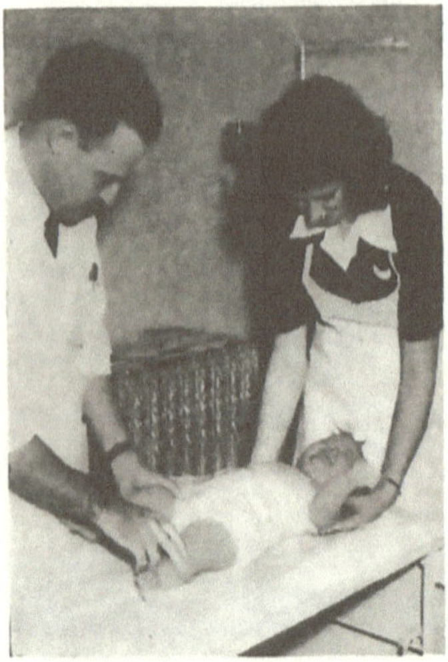

Special clinics for crippled children help many young North Carolinians grow up straight and strong.

result of certain jobs and certain working conditions. The Division of Industrial Hygiene checks on conditions to make sure that workers are protected from occupational diseases: the dust that may be present in textile mills, mining, lumber mills, and other industries; the injurious fumes of many chemical processes; lead poisoning and welding fumes in shipyards, and other such diseases. The Division also examines many factory workers for tuberculosis, making it possible for people to receive treatment before the disease has developed.

With these various divisions the State Board of Health, working mainly through county and city health departments, has done a great deal to make North Carolina a healthier place in which to live. But the Board, without the active help of the people, is not able to do everything. There are things which the people themselves must do.

Some of the diseases which used to be common, in North Carolina and all over the nation, can now be controlled. The people must do their part in helping to control them. With discoveries made by modern medical research it is possible to prevent epidemics of typhoid fever, smallpox, diphtheria, measles, and whooping cough, all of which caused a number of deaths in the past. The death rate from these diseases has become lower and lower every year as people learned how to prevent them.

Some problems of health—especially hookworm, malaria, and pellagra—have been more common in southern states than in any others. Through health education, hookworm has been practically wiped out, and malaria and pellagra have decreased greatly. The people themselves have done a great deal to help get rid of these diseases. But they are all diseases which may be common again in North Carolina if we are careless.

Hookworm is closely connected with a warm damp climate. It may be found where sanitary conditions are poor, where there are certain kinds of soil and a large amount of rain. It may be found where the temperature in winter is not cold enough to kill the hookworm larvae in the ground, and where the summers are so hot that children run around barefooted, giving the larvae a chance to burrow under the skin.

Since 1909 a great deal has been done to wipe out hookworm in every part of the South. Doctors discovered what medicine would cure people with hookworm. Sanitary conditions around homes and schools have been improved. And the causes of hookworm have been explained. Cure and prevention have done the job.

Malaria is found where the temperature is hot and where the land is swampy and low, like the coast of North Carolina. Swamps and lowlands, where water stands motionless, are breeding places for mosquitoes, which lay their eggs on stagnant water. It is a certain kind of mosquito which carries the malaria germ and gives the disease to the persons it bites.

A drive to wipe out malaria in the South has been going on for a number of years. There are special medicines for people who already have malaria. And everyone is urged to help get rid of mosquitoes. Doorways and windows in many houses now have screens to keep out mosquitoes, flies, and other insects. Mosquitoes that manage to get inside the houses should be killed with poison sprays.

Trees and plants along the edges of quiet ponds and creeks should be cut down. And drainage ditches should be kept clean. Mosquitoes lay their eggs in such places.

There is no connection between pellagra and climate, as there is between hookworm and climate and malaria and climate. Pellagra appears wherever people do not eat enough fresh protein food—fresh meat, and milk and other dairy products. Two hundred years or more before doctors knew the cause of pellagra, poor people in France and Spain were dying from it. It has been found in various parts of the United States.

Pellagra was once common in the South because many southerners ate salt meat, cornmeal, and molasses and very little else. These three foods were often produced at home on the farms. Or if they had to be bought, they were about the cheapest food there was. So "meat, meal, and molasses" became an eating habit for many southern people with low incomes. And the number of deaths from pellagra was high in the South.

Then in 1914 a doctor working for the United States Public Health Service set out to find the cause of pellagra. He discovered that it was caused by little or no fresh

Drinking water in towns and cities is tested regularly by trained specialists. The State Board of Health has a laboratory to which water from country wells and springs may be sent for testing.

protein food in the diet. And he found that if the disease had not developed too far, people could be cured by eating fresh meat and milk and other foods rich in proteins.

A campaign was started to tell people the cause and the cure for pellagra. The success of the campaign is seen in the decreasing number of deaths caused by this disease. In 1914, 24 out of every 100,000 people in North Carolina died of pellagra. Ten years later the number dropped to 10 out of every 100,000 people.

Then why worry about pellagra any more? Because the death rate from pellagra goes up and down with the income of the people. In the 1930's there were hard times. People lost their jobs or earned very low wages, and they often had no money for fresh meat and milk. In 1930 more than a thousand people died of pellagra in North Carolina alone. The number of pellagra deaths went up to 32 for every 100,000 people in the state. The death rate was higher than ever.

In 1940, jobs were more plentiful and people were making more money. They could afford to buy fresh meat and milk and other dairy products. And in 1940 the death rate from pellagra in North Carolina dropped to 5 in every 100,000.

A great deal has been found out about food since the days when the cause of pellagra was discovered. Scientists in nutrition have studied the effect of different foods on a person's health. According to these scientists, it is no longer enough for people to have plenty to eat. For people to be healthy they must have the right things to eat, the right kind of food.

To help us plan our diets, specialists have divided food into seven large general groups: (1) leafy, green, and yellow vegetables, (2) citrus fruits, tomatoes, raw cabbage, (3) potatoes and other vegetables and fruits, (4) milk, cheese, and ice cream, (5) meat, poultry, fish, eggs, dried peas and beans, (6) bread, flour, cereals, (7) butter or fortified margarine. These groups of food are often called the Basic Seven because they are essential to a healthy diet. Scientists tell us that every day children and grown people need foods in each group of the Basic Seven.

To be the right kind of food for people, fruits and vegetables have to be well fed themselves. They must be grown in soils which are well supplied with the necessary minerals. If soil lacks certain minerals, whatever is grown in the soil lacks these minerals, too. So to be well-fed today we must have enough food, the right kind of food, and well-fed food.

People are learning what they should eat and how to plan well-balanced meals. They learn at school, through home demonstration agents, through study groups, through articles in magazines and newspapers, and through books.

Public and private agencies have done a great deal to improve health in North Carolina. Yet records show that the state's rank in the nation is still low. To fill the needs of the people North Carolina must have more doctors, more dentists, more trained nurses, and more hospitals. And these must be scattered over the state, so that medical care will be within reach of all the people.

People with low incomes should be able to have medical care when they need it. Farm families and people in cities, towns, and villages must have proper food and the right kind of diet. Then more people will have a chance to grow

State and county agencies arrange for dentists to examine the teeth of children in the schools.

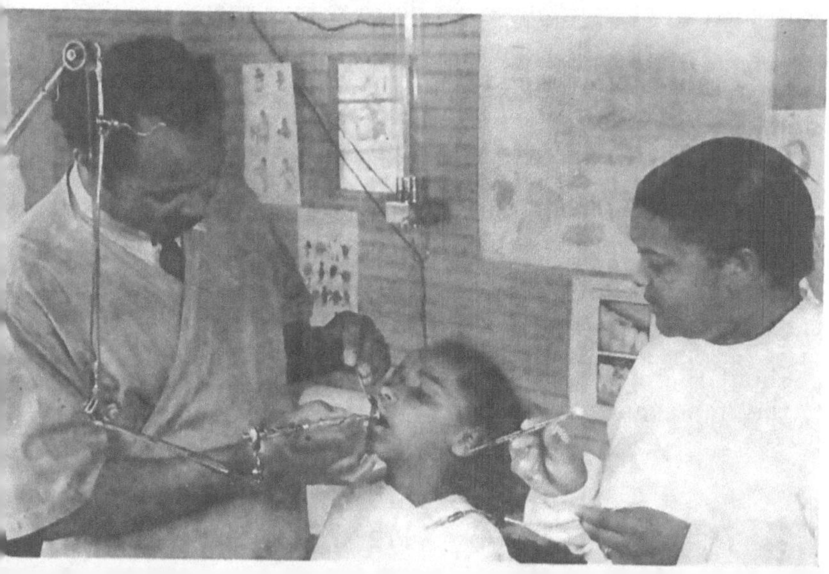

up strong and healthy, with less sickness and less need for medical care. Many North Carolinians think that better health is one of our most important needs. Because citizens of the state believe this, their representatives in the 1947 General Assembly passed a bill which will bring more hospitals and better medical care within reach of everyone.

As our agricultural and industrial wealth increases, our institutional wealth increases. And North Carolina has more resources for a healthier state.

Do You Know the Meaning of These Terms?

oral hygiene, vital statistics, epidemic, epidemiology, vaccines, serums, sanitation, pasteurized

Why Not Do These Things

1. Find out what are the most frequent causes of absences in your school. You will probably need the help of the principal or someone in his office in getting this information. After learning the chief causes of absences, discuss what your class can do to reduce them.

2. Arrange for a committee from your class to visit the County Board of Health and report about its work.

3. Invite the health officer to speak to your group about health conditions in your county, and about what is being done to improve them, what else needs to be done, and how you can help.

4. Consult your health book for further information about the diseases discussed in this chapter and about the importance of proper diet. If possible, arrange to

show film strips or movies dealing with causes and treatment of the diseases mentioned. Write to the State Board of Health for a list.

5. Have a series of class reports about the health and safety activities of the Good Health Association, the Red Cross, the Tuberculosis Association, Medical Care Commission, and any other organizations at work in North Carolina to make people healthier. What others are there?

6. Write to the Highway Safety Division of the State Highway Commission in Raleigh for material dealing with safety.

7. Read the interesting pamphlets *Public Health in America*, by Edgerton, and *Wanted: A Healthy South*, by Dickens.

8. Make a study of the winter gardens represented in your school grade. Could you grow other vegetables? Do you have an asparagus bed? Strawberries for the early spring? Which foods could be grown in your garden soil each month of the year?

47. *Recreation*

THE need for recreation is nothing new in the world. For a long time people have known that "All work and no play makes Jack a dull boy." But for a long time people had to spend all their energy making a living, and they had little chance for recreation.

Machines have changed this. They now do some of the work that men and women used to do, and they do it quickly. So as a rule, men no longer have to spend all their time and energy making a living. They have time for other things.

Factory workers are now at their machines only part of the day, and many of them work less than six days a week. The farmers who use machines to cultivate and harvest their crops can do their work in much less time than was necessary in the days of their grandfathers. Women who take care of their homes and cook the meals for their families often have many kinds of help: prepared foods; mechanical stoves, refrigerators, and cleaners; various appliances that make housework easier than it used to be. There are not so many little jobs for the children in a family to help with, especially if the family lives in a town or city.

This all means that there are a great many people—men, women, and children—who have time to do other things besides work. They have time for recreation, time

There are warm sunny days even during the winter when people can go out for a game of golf.

to do things which are restful and pleasant and different from their regular jobs.

Many people these days get recreation by taking part in sports. It may be a sport in which teams play against each other, like baseball or football. Or it may be a sport in which each man plays for himself, like golf and sometimes tennis. Games are recreation when people play for fun. But when men make their living by playing a game, baseball or football or some other sport, then the game is work. The players do something else for recreation.

Some people get their recreation doing things by themselves, enjoying one of the many hobbies that are popular today. One person collects stamps, another collects matchbook covers, another collects autographs. Gardening, carpentering, wood-carving, reading, weaving, painting —these are all recreations for someone.

Playing a musical instrument may be recreation. Camping, hiking, swimming, boating, hunting, or fishing may be recreation. Whatever a person does for a pleasant change and a rest is recreation.

Sometimes people find recreation in watching others do things. They are spectators. They go to theaters and motion pictures. Instead of playing a musical instrument, they like to listen while others play. Sports are recreation for them, but not as players. They go to baseball or football games because they enjoy watching the teams play.

North Carolina has many resources for recreation. Two of the greatest natural resources are probably the state's climate and scenery. There are mild winters, with many sunny days when people can play golf or tennis, go hunting or horseback riding. In the summer, when it is hot, the Carolina mountains are cool and beautiful. Or there are salty ocean breezes and fishing and many water sports along the coast. For the people who enjoy hiking and camping there are trails through mountain forests. For those who like boating and swimming there are lakes and rivers and the wide sounds of the Tidewater. For those who want to study trees or plants, birds or animals, there is a variety of plantlife and wildlife.

Sportsmen find good hunting and fishing in North Carolina, and many different agencies are helping to keep fish and game plentiful. Game laws allow hunting and fishing only in certain seasons and places, and limit the size and number of the fish or game a sportsman may take home. The work of other agencies helps, too. When the Soil Conservation Service and the TVA help farmers control soil erosion on their land, the soil is kept out of the streams and rivers. Then fish are more plentiful. When the Forest Service protects trees from fires, it is also helping sportsmen. A burning forest kills wildlife.

Nature has given North Carolina many resources for recreation, and the people who live here have these advantages close by, sometimes at their doorsteps. People from other states have heard so much about all the things to do and see in North Carolina that every year a large number of visitors come here for their vacations. The money they spend adds a good deal to the income of the state. Many people are able to earn their living in the hotels and camps, the summer resorts and the winter resorts where these visitors stay. Recreational resources mean dollars and cents, as well as pleasure, to many Carolinians.

Some of our towns and cities have men and women who are recreation directors. They help to plan different kinds of recreation for their communities. There are many neighborhood playgrounds. Boy Scouts, Girl Scouts, the YMCA, the YWCA, and many other groups have special programs of recreation. Factory workers sometimes have their swimming pools, club houses, and parks. They have their own teams on which they play basketball and baseball.

Playing the part of a squaw or an Indian brave in the school pageant may take a lot of work, but it can be a lot of fun.

Campers near lakes and rivers have various kinds of water sports.

The state has many private clubs where members play games and find various kinds of recreation. There are commercial places of amusement. And there are parks and athletic fields and other places which are managed by various agencies of the people.

The school is one of the government agencies helping people learn how to take part in various kinds of recreation. In the schools these days a student learns how to do a number of different things besides reading, writing, and arithmetic, and he often discovers something which will be a hobby for the rest of his life. Children in school learn to play many kinds of games. And they learn how to make and do many things—to paint and draw, to sing and dance, to play musical instruments. Some take parts in plays and pageants and debates. When children finish school, they often continue to enjoy the sports or other pastimes they learned in the classrooms or on the athletic fields at school.

In the libraries of cities, towns, and counties are shelves and shelves full of interesting books, where many people discover that reading can be one of the most exciting forms of recreation. Whatever a person is interested in, he can read about it in a book: exploration and travel, doing things and making things, ideas, people, places, gardening, stamp collecting, rocks, birds, animals.

The North Carolina Symphony Orchestra, which receives a small amount of money every year from the General Assembly, plays concerts all over the state. It gives many North Carolina musicians a chance to play in a large orchestra. And it makes it possible for many grown people and children to hear the music of great composers.

The state owns several lakes—six for white people and two for Negroes—where there are swimming, fishing, boating, and other water sports. In the state-owned parks people find various kinds of outdoor-fun: camping, picnics, trails for hikers, paths for horseback riders.

North Carolina also has national forests with places especially prepared for campers. The Croatan National Forest is near the coast. Pisgah Forest is in the west. Nantahala is in the southwest. And in the south central part of the state is the Uharie Purchase Unit.

Who doesn't enjoy a hike and a picnic supper?

The Great Smoky Mountains National Park, with almost half a million acres of land, lies partly in North Carolina and partly in Tennessee. Most of the Great Smoky Mountains are within the borders of this Park. For those who enjoy hiking, the Appalachian Trail winds through the Park along the North Carolina–Tennessee boundary. The Fontana Dam of the Tennessee Valley Authority is near the Park. The Dam itself is a sight to see. It is the highest dam in eastern America, one of the highest in the world. Around the shores of Fontana Lake and Hiwassee Reservoir and other large bodies of water, people may go boating, swimming, and fishing, or play outdoor games.

Parts of western Carolina are wild country, with many dense forests and with over one hundred mountain peaks rising a mile or more into the sky. In early summer the mountains are bright with rhododendron in bloom. In the fall they are bright with the turning leaves. And all year round there are the mountains themselves, sometimes clear against blue sky and sometimes lost in the mist that gives the Smokies their name.

The people of the state are beginning to realize that recreation is an important part of life today. They know they already have many recreational resources, and they want to make the most of them. Because of this, in 1945 the General Assembly established the North Carolina Recreation Commission, the first state commission of its kind in the nation. This Commission is making a study of what the state already has and what it needs. And it works with various other agencies of North Carolina in helping communities make plans to bring recreation within reach of everyone—white and Negro, young and old, the people on farms and in villages, towns, and cities.

It Will Be Fun

1. To have a hobby week when members of your class tell each other about their hobbies and exhibit their collections.

2. To sponsor the showing in your school or town of the film, *North Carolina Variety Vacationland.* This can probably be secured from the Department of Conservation and Development in Raleigh or from the University of North Carolina.

It Will Be Helpful

1. To make a recreational map of your state or county that might serve as a guide for visitors. Include national and state parks and forests.

2. To make a two-column chart. Show in one column what recreational opportunities are offered in your community now, and in the other column show what additional opportunities are needed to provide wholesome recreation for old and young.

3. To discuss and draw up a set of rules to guide a young person in planning a balanced program of recreation for himself.

Now Is the Time!

To Reread

1. Chapter 2, "The Wealth of the World."

2. Chapter 14, "Institutional Wealth."

3. Chapter 15, "The Wealth of North Carolina."

To Think About and Discuss

1. How institutions hold the key that unlocks the doors to greater natural, human, technological, and capital wealth in North Carolina.

2. How all the institutions discussed in this book depend upon one another for success.

3. How our institutions are the inventions of people who have worked together in groups.

4. How institutions extend the reach and multiply the powers of individuals and groups.

VI. NORTH CAROLINA TOMORROW

City and country are within easy reach of each other in North Carolina.

48. *A Balanced State*

THE people of North Carolina, as they look around the state and see what progress is being made, find they have many things to be proud of.

Farmers are growing valuable cash crops, and they are beginning to practice the diversified farming which will mean a better-balanced agriculture and a better use of the state's natural resources. Tobacco is now the leading crop in North Carolina, which raises more of it than any other state in the country. Cotton is still being grown. And besides tobacco and cotton, farmers are finding that their soils and climate will grow other things which are important sources of agricultural wealth. Peanuts and sweet potatoes are the main crops in some sections of the state. Fresh vegetables and fruits are grown in others. The value of livestock is gradually increasing as North Carolina farmers improve their pastures and buy better animals and more of them.

Industrial resources are growing. The textile mills, cigarette factories, furniture factories, and other industries already employ many workers. And with the state's wealth in power and other resources, North Carolina has possibilities for much greater industrial wealth.

As industries in North Carolina develop, they create jobs for many people besides those actually working at the machines. This is true no matter what raw materials are used or what the finished product happens to be.

From the production of cigarettes it is possible to get a general idea of how many people can be affected by one industry. It is a complicated story. Before it is ended, it has reached all over the nation and the world, touching the lives of millions and millions of people.

To begin with, the tobacco itself must be raised. This means jobs for all the people in any way connected with growing the crop and getting it to the factory. It means work for the farmer and for his wife and children. It means business for the company from which he buys his seed and fertilizer and for the stores where he gets his groceries and other supplies. It means work for the people who grow the seed and prepare the fertilizer and whatever else the farmer buys. It means work for all the men in the warehouse where he sells his crop: the auctioneers and buyers, the bookkeepers and clerks, the men who wheel the tobacco away after it has been sold, and the men who sweep the floors and help load the trucks.

Before the tobacco can be taken to the factory for manufacture, the factory must be standing there, ready to change the raw material into the finished product. There had to be workmen to build the factory, workmen to make the brick and steel or whatever materials the factory is built of, workmen to make the cigarette machines, workmen to build the factories in which the cigarette machines were manufactured, workmen to produce the raw materials from which the machines were made.

So far, this sounds something like a modern House that Jack Built, and in a way it is. And this is only the beginning. There must be men to produce the other necessities for manufacture—cigarette paper, tin foil, paper for packages, cellophane, cardboard cartons, boxes for shipping, to name only a few. And men must produce the raw materials that go into all these things, and build the factories where they are made, and make the machines of production.

There must be workers producing power for all the machines—men digging coal to be converted into steam or electricity, men running hydro-electric plants, men putting up lines to carry electric power through the country. And behind these men are men who planned the dams on the rivers to produce power, and men who built the dams, and men who made all the machines and tools for building.

Men must make automobiles and trucks and railway cars and ships for hauling goods, and docks and roads and railroads, and warehouses for storage. And there must be men to operate all these services.

There must be workers in banks so that tobacco, cigarettes, and other things may be bought or sold easily.

There must be men and women in the post offices and telephone companies and telegraph offices. These workers make it possible for people to do business by regular mail and air mail, by telephone and telegraph and cable, sending messages all over the world.

There must be clerks and stenographers and secretaries and bookkeepers and accountants. And there must be elevator operators and janitors and cleaning women and window washers. And there must be factories making typewriters, adding machines, elevators, and all the other necessary equipment.

Men and women in factories and offices, in transportation and communication and other services—the list continues almost endlessly. And as long as the list already

is, it could go on and on, and it might never be quite complete. Because for every worker already mentioned there are other people producing goods and services. People build the home where each worker lives. People grow the food he eats and make the clothes he wears. There are the doctors and dentists and lawyers he consults. There are the people who teach his children. There are people who make the books and newspapers he reads, play the music he hears, produce the plays and motion pictures he sees. And even this is not the end. Each person who performs a service for a worker has other people doing things for him.

It is a vast network—all the people everywhere who are helped to make a living because farmers in North Carolina grow tobacco and factories in North Carolina manufacture cigarettes. And a similar network lies behind every manufactured product from a large complicated machine to a spool of white cotton thread.

As long as the manufacture of cigarettes continues, it is the means by which numberless people make either part or all of their living. But if anything happens to reduce or stop the manufacture of cigarettes, the change is felt all through the network of production. Many people earn less money. Others lose their jobs.

The network of production is of great importance in the future development of North Carolina's resources. This network is one reason why experts advise people to diversify industry in the state just as they must diversify agriculture. Careful diversification will mean a balanced state.

If the network of production is made up of only one or two or three major industries, and if anything happens to one of them, the whole network feels the effect. Then a great many people in North Carolina have a hard time making a living. But if the network of production is made up of a number of industries, and if anything happens to one of those industries, then the others may keep the network from feeling the effect very much.

North Carolina's new agriculture and new industry must be diversified, and they must be diversified in ways which mean better use of resources and a better income for the people.

In spite of North Carolina's progress, the average income of the people is low. In 1946 the state's Department of Tax Research issued a report which showed that many people were having to live on very small incomes. In 1940, according to this report, North Carolina ranked forty-second among the states in average income for the whole population. In 1944 North Carolina ranked forty-fourth in average income.

The people of North Carolina will be able to earn a better living when agriculture and industry have been developed into a well-balanced, carefully planned use of the state's resources. In this development agriculture and industry are closely related. What helps one helps the other. One cannot thrive and grow strong if the other remains poor and weak.

Agriculture must have markets for its crops—people to buy the food that is raised, factories to buy the cotton and tobacco and other raw materials for manufacturing. And industry must have markets for its products—people to buy manufactured goods.

A farmer generally wants and needs a number of things he cannot produce for himself at home. But in order for him to be able to buy these things, he must be paid enough for whatever he grows and sells. How much he can buy depends on how much money he earns himself. If his income is low, his buying power is low.

Most of the workers in industry and the other people living in towns and cities who cannot raise their own food need the farmer's products. They would starve without his crops. But before these people can pay the farmer well for what he grows, they themselves must be paid enough for their own work. How much they can buy and how much they can pay for it depend on how much money they earn. If their incomes are low, their buying power is also low. This is true of everyone—farmers, industrial workers, clerks, business and professional workers, everyone.

A good income for the farmer means that he can buy things. A good income for the industrial worker means that he can buy things. And when people buy, they are helping to give jobs to others, making it possible for others to buy things, too. A well-developed agriculture and a well-developed industry will do much to increase the buying power of the people of North Carolina.

North Carolina, like the rest of the South, has always been a rural state. Long after other parts of the nation began developing their industries, the South remained an agricultural region. And in its agriculture North Carolina, like the South, has specialized in a few crops.

Since 1900, North Carolina has developed its industries until it ranks about twelfth in the nation as a manufacturing state. A great deal of wealth is created, and a great many people make their living in North Carolina industries. But like agriculture in North Carolina, these industries specialize in a few products.

As North Carolina's resources are put to wider use, a better-balanced agriculture and a better-balanced industry will play important parts in the development of a balanced state.

Can You

1. Draw pictures or cartoons which represent the "network of production" described in this chapter?

2. Describe a "network of production" that centers around another industry—furniture, for instance?

3. Prepare a booklet containing information and pictures which show how North Carolina can become a balanced state, and how all the people in the state will be benefited by her balanced development?

49. *North Carolina in the World Today*

EVEN with much thought and careful study, it is not easy for people to know how to make plans for the future. The world has a way of changing overnight, and everything sometimes has to change with it—people's plans along with everything else.

Many of the machines men have invented to help them make a living have also helped to change the world men live in. When machines do the work, men may often be out of jobs. Processes in manufacturing are done mechanically, and one man, operating a machine, can produce the amount of goods that once required the work of many men.

In textiles, for example, machines have gradually taken over the work that was once done by hand. A modern weaving room, with twenty-five weavers and two hundred power-operated looms, is said to produce an amount of woven goods that would require two thousand people if the work were done by hand. And the same kind of change has taken place in other industries.

When a machine is invented to do a difficult job, machinery may become more important in manufacturing than the individual worker. Even a skilled workman may find that his place in a factory is taken by a machine which can be operated by someone with little training or skill.

Man's inventions have also made the world much smaller than it used to be—smaller not in size, of course, but in terms of man's nearness to his neighbors. With today's communication, whatever happens anywhere may find its way into the daily radio programs and the daily newspapers. With today's transportation, the distance across the continent or around the world is measured in the speed of modern trains and ships and planes. The other side of the globe is always moving nearer.

In the world today, a man's next-door neighbor may live miles away, but he is within instant speaking distance by telephone and he can be reached within a few minutes by automobile. Ideas can now get from one place to another very quickly, and so can people and raw materials and manufactured goods.

People in North Carolina, like people everywhere, are realizing more and more clearly that the world today is indeed a small one. Whatever happens in another part of the world, either in the next state or on the other side of the globe, may have a close connection with what happens in North Carolina. The state cannot be separated from what goes on around it.

North Carolina is a part of the great Southeast—along with Virginia, South Carolina, Georgia, Florida, Kentucky, Tennessee, Alabama, Louisiana, Arkansas, and Mississippi. These eleven states of the southeastern region have much in common. Their people are much alike. Natural resources of soil and climate have made them primarily agricultural. Industrial resources lie close at hand, ready to be developed. There is great wealth of human resources—plenty of people to work. But the

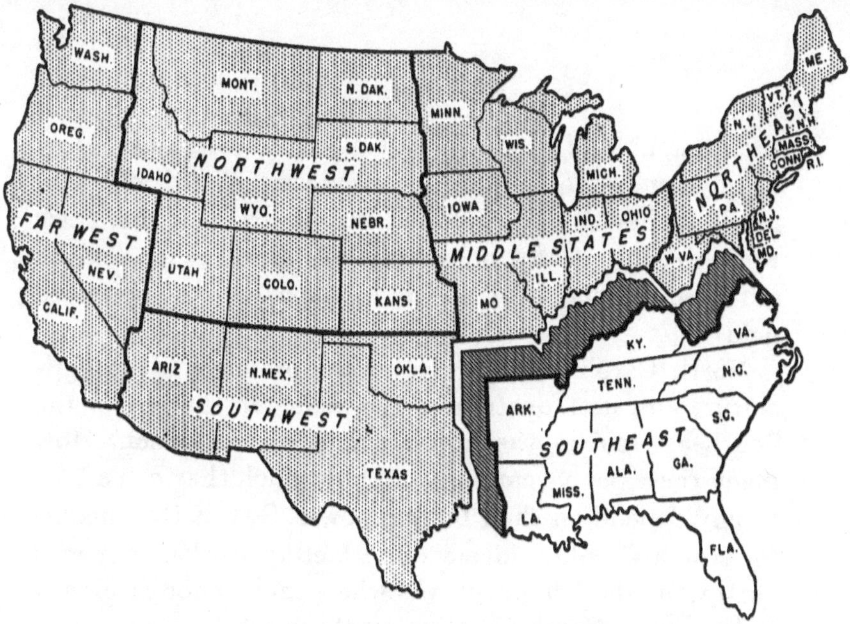

THE SOUTHEAST—ONE OF SIX REGIONS IN THE UNITED STATES

average income of many of these people is low. They do not make a very good living.

With similar conditions and similar problems, the states of the Southeast all feel the effect of what happens in any one of them. Tobacco and cotton farmers of North Carolina find that the price they get for their crops depends partly on the crops in the rest of the South. The same thing is true of other agricultural products. Workers in North Carolina textile mills find that their wages depend partly on how much textile workers in other southern states are paid. And the same thing is true of other manufactured products.

The Southeast is only one of the regions in the United States. There are five other regions: the Northeast, the Middle States, the Southwest, the Northwest, and the Far West. In each of them the people and their way of life are generally much alike.

What goes on in these other regions has an effect on the people of North Carolina. Some of these regions have the same kind of products that North Carolina has. How much money North Carolina products sell for and how much money North Carolina people are paid for their work depend partly on how much these other regions sell their products for and how much their workers are paid.

North Carolina sells some of the products of its agriculture and its industries to the people in the rest of the Southeast and in the other regions of the nation. How much these people can buy and how much they can afford to pay depend on their buying power. So a better income for North Carolina farmers and factory workers depends partly on whether people elsewhere make enough money to buy what North Carolina wants to sell.

North Carolina's close connection with other people does not stop with the Southeast or with the nation. It reaches across oceans into many other regions and nations. The network of production draws the whole world together. What North Carolina sells, what income North Carolina people earn, what kind of life North Carolina people live—all partly depend on what the rest of the world is making and selling and how it is living.

Long ago, in the first few years of the 1700's, John Lawson traveled through Carolina and wrote down his impressions of what he saw. The wealth of the land lay about him. "A delicious Country," he called it, with "sweet Air, moderate Climate and fertile Soil. . . . These are the Blessings that spin out the Thread of Life to its utmost Extent, and crown our Days with the Sweets of Health and Plenty, which, when joined with Content, render the Possessors the Happiest Race of Men upon Earth."

NORTH CAROLINA......
A PART OF THE SOUTHEAST

When these words were written, people were clinging to settlements scattered here and there along the Atlantic coast, and westward beyond their small clearings the land disappeared into the distance. How far did it go? No one knew.

Wondering about this unknown land and all the things that were perhaps waiting there, Lawson wrote that he could not even guess "as to the other Advantages the Country affords." From Carolina, he said, will probably "spring Productions that this Age does not dream of, and of much more advantage to the Inhabitants than any things we are yet acquainted with."

And Lawson was right. Many things which his own age never dreamed of have come out of Carolina. Looking back, we are able to see this.

Looking ahead, we cannot know, any better than could John Lawson, what the world will be like in a hundred years, or in ten years, or even in one. Things undreamed of may be just around the corner. But of one thing we can be certain: the world will change. And of one thing we can be sure: how the world changes will depend on all the people in it. People of every race and every color, people of every nation and every religion—in one way or another everyone will have some share in shaping the world of tomorrow.

Now Is the Time!

To Look Backward

1. What new words have you learned?

2. What are the most surprising facts you have learned about North Carolina and North Carolinians?

3. What would you consider the twenty most important facts to remember from your study of *North Carolina Today*?

4. What examples have you found of our similarity to other states in the Southeast? Of our dependence upon our neighbor states? Of their dependence upon us?

5. What examples have you found of our dependence upon other states and countries? Of their dependence upon us?

To Share With Others Some of the Things You Have Learned

1. Place in your permanent school museum everything of lasting value which you have made or collected while studying about the five kinds of wealth of North Carolina. Be sure everything is properly identified and meaningfully arranged, and that provision is made for its upkeep next year.

2. Write and produce a play or pageant that sums up what you have learned about North Carolina's wealth and how it can be increased. Invite not only your schoolmates but the public to attend this performance.

3. Make a collection of the best poems, stories, editorials, etc., written by your group during the year in connection with the study of North Carolina. Bind them in an attractive volume and leave them in your school or local library. You might like to illustrate these volumes. Why not try to work out a poem and a picture at the end of the volume which expresses your own idea of what North Carolina can and should become in the future? Maybe you can get an idea by reading aloud several times all the stanzas of Katherine Lee Bates, "America the Beautiful."

4. Sponsor the showing of films and strips about the Southeast, and the other regions of the United States. Order these from the University of North Carolina Bureau of Audio-Visual Aids, Chapel Hill. Perhaps you can see the film *Meet North Carolina*, which may be available through your nearest Standard Oil dealer.

To Look Ahead

1. Where and how can you start to do your part in increasing the wealth of North Carolina?

2. What occupation can you prepare to enter that will benefit both you and the state?

3. What courses in high school can you take which will help in your preparation for this occupation?

4. What habits and attitudes do you need to develop in order to succeed in a North Carolina which must become more and more a part of "One World"?

References for *Aids to Learning*

BOOKS

THE books on the Free List of the State Board of Education are marked [F]; those on the Supplementary List are marked [S]. These are available from the Division of Textbooks, State Board of Education, Raleigh. Many of the other titles are listed in the Library Book Catalogue of the North Carolina Education Association, Raleigh. When the books are not in school or local libraries and cannot be purchased, they may in most cases be borrowed by the teacher from the University Extension Library at Chapel Hill. Or they may be secured through the local librarian from the North Carolina Library Commission, Raleigh.

ALLEE, MARJORIE HILL. *Road to Carolina.* Boston: Houghton Mifflin, 1932.

———. *Susanna and Tristam.* Boston: Houghton Mifflin, 1929.

ALLEN, ADAM. *Dynamo Farm.* Philadelphia: Lippincott, 1942.

[S] ARNETT, ALEX MATTHEWS. *The Story of North Carolina.* Chapel Hill: University of North Carolina Press, 1942. 80¢.

BARNES, RUTH, comp. *I Hear America Singing.* Philadelphia: Winston, 1937.

BOYD, JAMES. *Drums.* New York: Scribner's, 1928.

BRAWLEY, BENJAMIN. *Negro Builders and Heroes.* Chapel Hill: University of North Carolina Press, 1937.

BRUERE, MARTHA BENSLEY. *Your Forests.* Philadelphia: Lippincott, 1945.

BRYSON, HERMAN J. *The Story of the Geologic Making of North Carolina.* Raleigh: Department of Conservation and Development, 1928. Free.

F CARROLL, FRANKLIN B. *Understanding Our World.* Book 2. Philadelphia: Winston, 1939.

S CHAPMAN, PAUL W. *Better Rural Careers.* Chicago: Science Research Associates, 1942. 83¢.

F ———. *Successful Farming in the South.* Atlanta: Turner E. Smith and Company, 1942. 86¢.

S CONNOR, R. D. W. *Makers of North Carolina History.* Raleigh: Alfred Williams, 1930. 75¢.

DEUSING, MURL. *Soil, Water, and Man.* Basic Social Education Series. Evanston: Row, Peterson, 1941. 32¢.

DICKENS, DOROTHY. *Wanted: A Healthy South.* Atlanta: Southern Regional Council, Inc., 1946. 25¢.

EATON, ALLEN H. *Handicrafts of the Southern Highlands.* New York: Russell Sage Foundation, 1937.

EDGERTON, AVIS E. *Public Health in America.* Evanston: Row, Peterson, 1941. 32¢.

FITZGERALD, MARY NEWMAN. *The Cherokee and His Smoky Mountain Legends.* Asheville: The Stephens Press, 1946. 50¢.

S FLYNN, HARRY E., AND PERKINS, FLOYD E. *Conservation of the Nation's Resources.* New York: Macmillan, 1941.

GAER, JOSEPH. *Fair and Warmer.* New York: Harcourt, Brace, 1939. 95¢.

GRAY, ELIZABETH JANET. *Jane Hope.* New York: Viking, 1934.

———. *Meggy McIntosh.* New York: Viking, 1930.

S GREEN, CHARLOTTE HILTON. *Birds of the South.* Chapel Hill: University of North Carolina Press, 1934.

S ———. *Trees of the South.* Chapel Hill: University of North Carolina Press, 1939.

GREEN, PAUL. *The Lost Colony.* Chapel Hill: University of North Carolina Press, 1937.

HAMLIN, C. H. *Ninety Bits of North Carolina Biography.* New Bern, N. C.: Owen G. Dunn, 1946.

F HAMMONDS, CARSIE, AND WOODS, RALPH H. *Today's Agriculture.* Philadelphia: Lippincott, 1938. 93¢.

HENDRICKS, W. C., ed. *Bundle of Troubles and Other Tar Heel Tales.* Durham: Duke University Press, 1943.

HOLMES, J. S. *Common Forest Trees of North Carolina.* Raleigh: Department of Conservation and Development, 1944. 10¢.

HUFF, DARRELL. *Twenty Careers of Tomorrow.* New York: McGraw-Hill, 1945.

KELLY, PAUL, AND BASKERVILLE, JOHN C. *The Story of Conservation in North Carolina.* Raleigh: Department of Conservation and Development, 1941. Out of print but available in many schools.

KEY, ALEXANDER. *With Daniel Boone on the Caroliny Trail.* Philadelphia: Winston, 1941.

KNOX, ROSE B. *Gray Caps.* Garden City, New York: Doubleday, Doran, 1937.

———. *Marty and Company.* Garden City, New York: The Sun Dial Press, 1942.

LANDIS, BENSON Y., ed. *Yearbook of American Churches.* Lebanon, Pa.: Sowers Printing Company, 1945 edition. Published biennially.

LATHROP, VIRGINIA TERRELL. *Educate a Woman: Fifty Years of Life at Woman's College of the University of North Carolina.* Chapel Hill: University of North Carolina Press, 1942.

LEFLER, HUGH T., ed. *North Carolina History Told by Contemporaries.* Chapel Hill, University of North Carolina Press, 1934.

LENSKI, LOIS. *Blue Ridge Billy.* Philadelphia: Lippincott, 1946.

LEYSON, BURR WATKINS. *Plastics in the World Tomorrow.* New York: Dutton, 1944.

LIDE, ALICE ALISON. *Johnny of the 4-H Club.* Boston: Little, Brown, 1941.

MEANS, FLORENCE C. *Shuttered Windows.* Boston: Houghton Mifflin, 1938.

MELROSE, MARY, AND OTHERS. *Nature's Bank—The Soil.* Washington: National Wildlife Federation, Inc., 1942.

MERSEREAU, SAMUEL FOSTER. *Materials of Industry: Their Distribution and Production.* New York: McGraw-Hill, 1941.

F NEWSOME, A. R., AND LEFLER, H. T. *The Growth of North Carolina.* Yonkers, New York: World Book Company, 1942.

North Carolina: A Guide to the Old North State. Federal Writers' Project. Chapel Hill: University of North Carolina Press, 1939.

North Carolina Agricultural Statistics. Raleigh, North Carolina: Crop Reporting Service, State Department of Agriculture. Issued annually.

North Carolina Manual. Raleigh: Secretary of State. Issued every two years. Single copies to schools free.

PARKER, BERTHA MORRIS. *Balance in Nature.* Basic Science Education Series. Evanston: Row, Peterson, 1941. 32¢.

———. *Soil.* Basic Science Education Series. Evanston: Row, Peterson, 1943. 32¢.

PARKER, WILLIS A. *Our Friendly Neighbors.* Asheville: The Stephens Press, 1945.

PEARSON, T. G., AND OTHERS. *Birds of North Carolina.* Raleigh: Department of Agriculture, 1942.

PICKWELL, GAYLE. *Weather.* New York: McGraw-Hill, 1938.

F POOLEY, ROBERT C., AND WALCOTT, FRED G. *Contact! Growth in Reading: Book Two.* Chicago: Scott, Foresman, 1942.

PORTER, ELLA WILLIAMS. *Sandra Kendall of the 4-H Club.* New York: Dodd, Mead, 1942.

REED, W. MAXWELL, AND BRONSON, WILFRED S. *The Sea for Sam.* New York: Harcourt, Brace, 1935.

ROGERS, FRANCES, AND BEARD, ALICE. *Fresh and Briny: The Story of Water as Friend and Foe.* New York: Stokes, 1936.

S ROWE, NELLIE M. *Discovering North Carolina.* Chapel Hill: University of North Carolina Press, 1940. 70¢

SALLEY, KATHERINE BATTS, ed. *Life at St. Mary's.* Chapel Hill: University of North Carolina Press, 1942.

SMITH, J. RUSSELL, AND PHILLIPS, M. OGDEN. *North America: Its People and the Resources, Development, and Prospects of the Continent as the Home of Man.* New York: Harcourt, Brace, 1940.

F SMITH, VICTOR C., AND TRAFTON, GILBERT H. *Enjoying Science.* Philadelphia: Lippincott, 1942.

STILLMAN, ALBERT. *Drums Beat in Old Carolina.* Philadelphia: Winston, 1939.

F WARREN, JULE B. *North Carolina Yesterday and Today.* Raleigh:

State Department of Public Instruction, 1942. On free list for fifth grade.

———. *The People Govern North Carolina.* Raleigh: The Warren Publishing Company, 1946.

F WATKINS, RALPH K., AND PERRY, WINIFRED. *Understanding Science.* New York: Macmillan, 1940.

WETHERINGTON, JULIA, AND OTHERS. *Science for the Elementary School.* Raleigh: issued by the State Superintendent of Public Instruction, 1941. Publication No. 227.

The World Almanac and Book of Facts. New York: New York World Telegram. Published yearly. 85¢.

WORTH, KATHRYN. *The Middle Button.* Garden City, New York: Doubleday, Doran, 1941.

———. *They Loved to Laugh.* Garden City, New York: Doubleday, Doran, 1942.

ZIM, HERBERT S., AND COOPER, ELIZABETH K. *Minerals: Their Identification, Uses, and How to Collect Them.* New York: Harcourt, Brace, 1943.

FILMS

The Conservation of Natural Resources. E.B.F.* (16 mm., sound. 10 minutes) Shows how water, forests, and land were once wasted but are now being better conserved. U.N.C.

*Food and Soil.** U.S.D.A. (16 mm., sound, in color. 10 minutes) Presents in an entertaining way the fact that food comes directly or indirectly from productive soil. Extension Service.

For Years to Come. U.S.D.A. (16 mm., sound, in color. 22 minutes) Shows the life of a farm family during the year in which they began better land use. Extension Service.

A Heritage We Guard. U.S.D.A. (16 mm. or 35 mm., sound. 30 minutes) Shows interrelationship of wildlife to conservation and gives the steps to take in order to restore and conserve natural resources. Extension Service.

* The Encyclopedia Britannica Films (E.B.F.) are available from the Bureau of Audio-Visual Aids, University of North Carolina, Chapel Hill. The Department of Agriculture films are available through local extension agents or the Agricultural Extension Service, State College, Raleigh. Some are handled by the University also.

Live At Home. U.S.D.A. (16 mm., sound, in color. 11 minutes) Shows how growing of food and many other products at home is both profitable and patriotic. Extension Service.

Meet North Carolina. (16 mm., sound, in color. 30 minutes) Presents North Carolina sports and resort areas, agriculture and manufacturing. Standard Oil Company.

North Carolina Variety Vacationland. (Conservation and Development Department) (16 mm., sound, color. 45 minutes) Covers the scenic and recreational resources of the state from the sea to the mountains. U.N.C.

Patterns of American Rural Art. U.S.D.A. (16 mm., sound, in color. 10 minutes) Portrays objects of rural handicraft exhibited in Rural Art Exhibit held in Washington in 1940. Extension Service or U.N.C.

The South Grows Green. U.S.D.A. (16 mm., sound, color. 25 minutes) Shows how crop diversification is benefiting the South in many ways. Extension Service.

The Southeastern States. E.B.F. (16 mm., sound. 10 minutes) Pictures ways of living in these states, and shows how this region is affected by the rest of the nation. Similar films about five other regions. U.N.C. Film strips on same subject are also available. E.B.F.

Selected References for the Teacher

IN addition to the material listed for students, the teacher will find the following references useful. They are especially chosen to give a general background of information about recent social and economic conditions in the state and, to a lesser degree, in the South. No works of history or historical fiction are included. A comprehensive list of books, pamphlets, and other materials may be found in a bulletin prepared by the State Department of Public Instruction: *Studying the State of North Carolina*, Publication No. 258, 1947. This bulletin may be obtained from the Department in Raleigh for twenty-five cents.

ALFORD, LEE F., ed. *The North Carolina Yearbook.* Raleigh: News and Observer. Published annually. $1.00. A directory of state and local officials, and a compilation of facts about farming, industry, taxes, education, etc.

BATHHURST, EFFIE G. *Conservation Excursions.* Bulletin 1939, No. 13. U. S. Office of Education. Washington: Government Printing Office, 1940. 15¢.

BULLOCK, BENJAMIN F. *Practical Farming for the South.* Chapel Hill: University of North Carolina Press, 1946. Discusses in non-technical fashion soil, plant and animal production, better land use, etc.

COUCH, W. T., ed. *Culture in the South.* Chapel Hill: University of North Carolina Press, 1934. Chapters on agriculture, industry, education, economics, literature, art, etc.

CRUIKSHANK, J. W. *North Carolina Forest Resources and Industries*. Miscellaneous Publication No. 533. Forest Service, U. S. Department of Agriculture. Washington: Government Printing Office, 1943. 25¢.

Directory of Manufacturing Firms in North Carolina. Raleigh: Department of Labor, 1944. $1.00. Includes information about each county.

EMBREE, EDWIN R. *American Negroes: A Handbook*. New York: John Day, 1942. $1.00.

GLESINGER, EGON. *A Post War Program for North Carolina's Forest Industries*. Bulletin No. 45. Raleigh: Department of Conservation and Development, 1944. Free.

HARRILL, L. R. *A Brief History and Summary of Thirty Years of 4-H Club Work in North Carolina*. Raleigh: North Carolina Extension Service, State College Station, 1939. Free.

HAYNES, WILLIAMS. *Southern Horizons*. New York: Van Nostrand, 1946. Lively chapters on cotton, wood pulp and other uses of our forest resources, water power, the oils we eat, etc.

HERRING, HARRIET L. *North Carolina's New Industrial Opportunity*. Raleigh: State Planning Board, 1945. Single copies free.

HOBBS, S. H., JR. *North Carolina: Economic and Social*. Chapel Hill: University of North Carolina Press, 1930. Available in many school libraries.

JOHNSON, GERALD W. *The Wasted Land*. Chapel Hill: University of North Carolina Press, 1938. A brief, challenging essay on the way the South has wasted both land and people.

KELLOGG, CHARLES E. *The Soils That Support Us*. New York: Macmillan, 1941.

KEPHART, HORACE. *Our Southern Highlanders*. New York: Macmillan, 1926.

LARKINS, JOHN R. *The Negro Population of North Carolina*. Raleigh: State Department of Public Welfare, 1945. Free.

LILIENTHAL, DAVID. *T V A: Democracy on the March*. New York: Harper, 1944. Also New York: Pocket Books. 25¢. A stirring description of the program and the philosophy of the T V A.

MCKIMMON, JANE S. *When We're Green We Grow*. Chapel Hill: University of North Carolina Press, 1945. Pioneering in home demonstration work in North Carolina.

NEWBOLD, N. C., ed. *Five North Carolina Negro Educators.* Chapel Hill: University of North Carolina Press, 1939. $1.00.

NIXON, H. C. *Forty Acres and Steel Mules.* Chapel Hill: University of North Carolina Press, 1938. Opportunities for the rural South.

North Carolina Basic County Data. 2 volumes. Raleigh: State Planning Board, 1946. In school and public libraries. Separate sheets for single counties are free to teachers.

ODUM, HOWARD W. *Southern Regions of the United States.* Chapel Hill: University of North Carolina Press, 1936.

PEATTIE, RODERICK, ed. *The Great Smokies and the Blue Ridge.* New York: Vanguard, 1943. Descriptions of the country, the people, and their crafts, customs, music, etc.

Population. First Series, Number of Inhabitants in North Carolina. Census Bureau, U. S. Department of Commerce. Washington: Government Printing Office, 1941. 10¢.

———. Second Series, Characteristics of Population, North Carolina. Census Bureau, U. S. Department of Commerce. Washington: Government Printing Office, 1941. 15¢.

RANDOLPH, E. E. *Suitability of North Carolina Trees for Chemical Conversion and Certain Other Uses.* Raleigh: North Carolina Experiment Station, State College, 1945. 50¢.

Resource-Industry Series. Raleigh: Department of Conservation and Development, 1946–1947. Single copies free.

SEARS, PAUL B. *Life and Environment.* New York: Teachers College, Columbia University, 1939. Discusses the interrelationship of all living things.

VANCE, RUPERT B. *All These People.* Chapel Hill: University of North Carolina Press, 1946. A study of the nation's population needs and natural resources in their relation to the people and the culture of the South.

———. *Human Geography of the South.* Chapel Hill: University of North Carolina Press, 1939. Available in many school libraries.

WAGER, PAUL W. *North Carolina: the State and its Government.* New York: Oxford Book Company, 1947.

The following publications will be sent to teachers free upon request.

Agricultural Review. Raleigh: State Department of Agriculture. Twice monthly.

Agricultural Statistics. Raleigh: Crop Reporting Service, State Department of Agriculture. Yearly.

Extension Farm-News. Raleigh: Agricultural Extension Service, N. C. State College. Monthly.

Health Bulletin. Raleigh: State Board of Health. Monthly.

News Letter. Chapel Hill: Extension Division, University of North Carolina. Twice monthly.

Public Welfare News. Raleigh: State Board of Public Welfare. Quarterly.

Research and Farming. Raleigh: Experiment Station, N. C. State College. Quarterly.

Resource-Use Education. Chapel Hill: Committee on Southern Regional Studies and Education. Quarterly.

State School Facts. Raleigh: Department of Public Instruction. Monthly.

The U.C.C. Quarterly. Raleigh: Unemployment Compensation Commission Informational Service. Quarterly.

Index

www.ingramcontent.com/pod-product-compliance
Lightning Source LLC
LaVergne TN
LVHW090259210926
848480LV00001B/77